Isaac Mathematics

Essential
pre-university mathematics for sciences

Julia Riley & Mark Warner
Cavendish Laboratory, University of Cambridge

Periphyseos Press
Cambridge, UK.

Co-published in Cambridge, United Kingdom, by
Periphyseos Press and Cambridge University Press.

www.periphyseos.org.uk and www.cambridge.org

Isaac Mathematics
Essential pre-university mathematics for sciences
Open Government Licence.

First published & First reprint 2018
Second reprint 2020
Co-published adaptation, 2020

Printed and bound in the UK by Short Run Press Limited, Exeter.

Typeset in LaTeX

A catalogue record for this publication is available from the British Library

ISBN 978-1-8382160-2-3 Paperback

Use this book in parallel with the electronic version at isaacmaths.org. Marking of answers and compilation of results is free on Isaac. Register as a student or as a teacher to gain full functionality and support.

Front cover image: An exponentially growing horn centred on a logarithmic spiral. (created with WolframMathematica®)

used with kind permission of M. J. Rutter.

Isaac Mathematics for Students and Teachers

Chapters 1–6 cover the Mathematics questions of levels 1–6 of the Isaac Sciences OPAL (Open Platform for Active Learning), corresponding to the last 3 years at school and to the foundation of university studies (1 – pre-A Level, 2–3 – AS, 2–5 – A level, 6 – further Maths and beyond). Chapter 7 addresses fundamental, perhaps unfamiliar, applications of mathematics that have been used in advanced Isaac events and for student extension.

The material of Chapters 1–6 is core to school mathematics courses, underpinning all A-level sciences and looking forward to university mathematics, sciences and engineering. It can be used by all students, even including those just doing mathematics without sciences.

All questions in this book can be answered on-line on Isaac at `isaacmaths.org` where there is immediate marking and feedback, with linked concept pages and several levels of hints. Many questions are straightforward, at least after practice(!), and are essential for the fluency and confidence required for creativity in higher maths and sciences. Other questions are challenging. In both cases, one learns by actively **doing**, hence the Isaac mantra "Put the mouse down and pick up the pencil". Working on paper is essential – answering on the OPAL is *not* the initial action.

Students should register (free) on Isaac since then progress is recorded (but remains private, except that set homework and shared with teachers). Indeed, problems done at any stage are declared as done, should they later be set as homework. Accomplishments on Isaac can be shared on CVs or for university applications.

Teachers registered on Isaac can apply for teacher status to set homework and have it marked, with the detailed results (question-by-question and student-by-student) instantly analysed for reporting.

JR & MW, Cambridge, 2018

Acknowledgements

We are grateful to Robin Hughes (RWH in Chapter 7) for inspiration and for contributing imaginative and challenging questions. Heather Peck helped greatly with the vectors, exponentials and calculus materials in Chapter 7. Gareth Conduit made a collection of integrals available to us.

We owe Dr Luciana Bonatto a huge debt for her work and creativity in bringing this book to fruition. She designed materials and oversaw the implementation of sketching sections and the extension of chapter 7. In presenting the book online, with its associated greater functionality, she was ably helped by Henry Boult, Josh Brown, Michael Conterio, Ben Hanson, Abigail Peake and James Sharkey.

Uncertainty and Significant Figures

In science, numbers represent values that have uncertainty and this is indicated by the number of significant figures in an answer.

Significant figures

When there is a decimal point (dp), all digits are significant, except leading (left-most) zeros: 2.00 (3 sf); 0.020 (2 sf); 200.1 (4 sf); 200.010 (6 sf).

Numbers without a dp can have an *absolute accuracy*: 4 people; 3 electrons.

Some numbers can be ambiguous: 200 could be 1, 2 or 3 sf (see below). Assume such numbers have the same number of sf as other numbers in the question.

Combining quantities

Multiplying or dividing numbers gives a result with a number of sf equal to that of the number with the smallest number of sf:

$x = 2.31, y = 4.921$ gives $xy = 11.4$ (3 sf, the same as x).

An absolutely accurate number multiplied in does not influence the above.

Standard form

On-line, and sometimes in texts, one uses a letter 'x' in place of a times sign and ^ denotes "to the power of":

1800000 could be 1.80x10^6 (3 sf) and 0.0000155 is 1.55x10^-5

(standardly, 1.80×10^6 and 1.55×10^{-5}).

The letter 'e' can denote "times 10 to the power of": 1.80e6 and 1.55e-5.

Significant figures in standard form

Standard form eliminates ambiguity: In $n.nnn \times 10^n$, the numbers before and after the decimal point are significant:

$191 = 1.91 \times 10^2$ (3 sf); 191 is $190 = 1.9 \times 10^2$ (2 sf); 191 is $200 = 2 \times 10^2$ (1 sf).

Answers to questions

Here, and on-line, give the appropriate number of sf: for example, when the least accurate data in a question is given to 3 significant figures, then the answer should be given to three significant figures; see above. Too many sf are meaningless; giving too few discards information. Exam boards also require consistency in sf.

Contents

Isaac Mathematics for Students and Teachers   i

Acknowledgements   i

Uncertainty and Significant Figures   ii

1   Level 1   **1**

    1.1   Algebraic Manipulation – rearranging equations, units   . . .   1

    1.2   Quadratic Equations – factorising, solving   3

    1.3   Simultaneous Equations – linear, quadratic   5

    1.4   Trigonometry – angles, triangles   7

    1.5   Functions – evaluating, transforming, sketching   9

    1.6   Graph Sketching – simple function types, trig functions   . . .   10

2   Level 2   **14**

    2.1   Algebraic Manipulation – inequalities, indices   14

    2.2   Trigonometry – sin, cos, tan, triangles   16

    2.3   Simple Shapes – area, volume   18

    2.4   Vectors – notation; adding, resolving components   20

    2.5   Functions – polynomials, symmetry; transforming   23

    2.6   Differentiation – powers, stationary points   26

    2.7   Graph Sketching – powers of x, polynomials   28

3   Level 3   **34**

    3.1   Trigonometry – circles, radians/degrees   34

    3.2   Functions – exponentials, logarithms   36

    3.3   Series – binomial expansion   38

    3.4   Differentiation – powers, stationary points   39

    3.5   Integration – powers, definite/indefinite integrals   43

    3.6   Graph Sketching – exponentials, logs   46

4   Level 4   **51**

4.1	Trigonometry – addition of angles formulae	51
4.2	Functions – e, ln, composite, modulus	53
4.3	Series – arithmetic, geometric, binomial	55
4.4	Differentiation – e, ln, trig, chain rule, product rule	57
4.5	Integration – e, ln, trig	59
4.6	Graph Sketching – summing functions; e and ln; modulus	61

5   Level 5   67

5.1	Vectors – scalar products	67
5.2	Functions – rational, polynomials	69
5.3	Differentiation – implicit, chain rule, product rule	70
5.4	Integration – by parts, substitution, trig identities	73
5.5	Differential Equations – first order	74
5.6	Graph Sketching – products of functions; algebraic functions	76
5.7	Advanced Algebraic Manipulation – on-line	81

6   Level 6   82

6.1	Vectors – vector products	82
6.2	Functions – hyperbolic, sinc	85
6.3	Series – Maclaurin, Taylor	91
6.4	Differentiation – inverse functions, chain rule, product rule	93
6.5	Integration – by parts, partial fractions, substitution	98
6.6	Differential Equations – first order, second order	99
6.7	Graph Sketching – rational, hyperbolic and other functions	101

7   Applications to Sciences   107

7.1	Advanced vectors – 1	107
7.2	Advanced vectors – 2	111
7.3	The calculus of change – Exponentials	116
7.4	Words to Physics to Calculus	124
7.5	The calculus of change – Population	128
7.6	Parametric curves, circular coordinates, and vector calculus	133
7.7	Wind–driven yachts, sand yachts and ice boats	141
7.8	Rays, rainbows, and caustics	147
7.9	Dr. Conduit's 101 Integrals	151

Level 1

> You might find it useful to look at the following on-line concept pages.
> - Algebraic Manipulation - Rearranging Equations and Units - Level 1:
>   isaacphysics.org/concepts/cm_algebra_manip

1. a) Rearrange the equation of motion $v = u + at$, where v is the speed of a particle which has been accelerating at a constant rate a for a time t from an initial speed u, to make t the subject.

   b) Rearrange the equation of motion $v = u + at$, this time to make a the subject.

   c) In the equation of motion for a uniformly accelerating body the distance s travelled in time t is given by $s = \frac{1}{2}(u + v)t$, where u and v are the initial and final speeds. Rearrange the equation to make u the subject.

   d) Looking again at $s = \frac{1}{2}(u + v)t$, rearrange the equation to make t the subject.

2. a) In the equation of motion for a body accelerating uniformly at a rate a, the distance s travelled in time t is given by $s = ut + \frac{1}{2}at^2$, where u is its initial speed. Rearrange the equation to find an expression for t assuming $u = 0$.

   b) Rearrange the equation $s = ut + \frac{1}{2}at^2$ again, this time without assuming $u = 0$, to make t the subject.

   c) In the equation of motion for a body accelerating uniformly at a rate a, the relationship between the distance travelled s and the initial and final speeds u and v is given by $v^2 = u^2 + 2as$. Rearrange the equation to find an expression for u.

   d) Rearrange $v^2 = u^2 + 2as$ again, to make s the subject.

3. a) Rearrange the equation $F = ma$, which relates the force F on a body to its mass m and acceleration a, to make a the subject of the equation.

   b) Rearrange the equation $W = mg$, which relates the weight W of a body to its mass m, to make m the subject of the equation.

4. a) Rearrange the equation $\rho = \frac{m}{V}$, relating the density ρ of a body to its mass m and volume V, to make m the subject of the equation.

   b) Rearrange $V = IR$, which relates the voltage V across a resistance R to the current I through it, to make I the subject of the equation.

   c) Considering again $V = IR$, make R the subject of the equation.

5. a) Rearrange $E_k = \frac{1}{2}mv^2$, which gives the kinetic energy E_k of a body of mass m travelling with speed v, to make v the subject of the equation.

   b) Rearrange $P = V^2/R$, which gives the power P dissipated in a resistance R when the voltage across it is V, to make V the subject.

6. Rearrange $F = GMm/r^2$, the expression for the gravitational force F between two masses M and m a distance r apart, to make r the subject.

7. a) The equation $F = ILB\sin\theta$ gives the force F on a length L of wire carrying a current I in a magnetic field B, when the magnetic field is at an angle of θ to the direction of current flow. Rearrange the equation to find an expression for θ.

   b) Rearrange $x = A\cos(2\pi ft + \phi)$, which gives the displacement x of an object oscillating at a frequency f with an amplitude A, to make t the subject.

8. Using $v = u + at$, find v if $u = 3.0 \text{ m s}^{-1}$, $a = 9.8 \text{ m s}^{-2}$ and $t = 2.0 \text{ s}$.

9. Using $v = u + at$, find v if $u = 3.0 \text{ cm s}^{-1}$, $a = 9.8 \text{ m s}^{-2}$ and $t = 2.0 \text{ ms}$.

10. Find the force, in newtons, on a body of mass 3.0 kg which is accelerating at 2.5 m s^{-2}.

In addition to the concepts from previous sections, see:
- Quadratic Equations - Level 1 and 2:
  `isaacphysics.org/concepts/cm_algebra_quadr`

1. Consider the equation $3b^2 - 2b - 1 = 0$.

   a) Factorise the left hand side of the equation.

   b) Give the exact value of the root closest to zero.

2. Consider the equation $9q^2 + 9q + 2 = 0$.

   a) Factorise the left hand side of the equation.

   b) Give the exact value of the root closest to zero.

3. Consider the equation $-2s^2 - 5s + 25 = 0$.

   a) Factorise the left hand side of the equation.

   b) Give the exact value of the root closest to zero.

4. Consider the equation $8s^2 + 2s - 6 = 0$.

   a) Factorise the left hand side of the equation.

   b) Give the exact value of the root closest to zero.

5. Consider the equation $k + 3 = \dfrac{1 - k}{k + 2}$.

   a) Rearrange the equation to give a quadratic equation in which the right hand side is zero. (As a first step, eliminate the fraction by multiplying through by an appropriate expression.)

   b) Factorise the left hand side of the equation derived in part a).

   c) Find the roots of the equation.

6. Solve the equation $3p^2 - 6p - 4 = 0$. What is the solution closest to zero? Give your answer to 3 sf.

7. Solve the equation $m^2 + 3m + 1 = 0$. What is the solution closest to zero? Give your answer to 3 sf.

8. Solve the equation $-4z^2 + z + 1 = 0$. What is the solution closest to zero? Give your answer to 3 sf.

9. Find the value of v closest to zero if $\dfrac{3-v}{1-3v} = \dfrac{2+v}{1+2v}$. Give your answer to 3 sf.

10. Show that the solution to the equation $mp^2 + bp + k = 0$ can be written as $p = -\gamma \pm \sqrt{\gamma^2 - \omega^2}$.

a) Hence find an expression for γ in terms of one or more of the constants m, b and k in the original equation.

b) Give also an expression for ω in terms of one or more of the constants m, b and k.

1.3   Simultaneous Equations – linear, quadratic

> In addition to the concepts from previous sections, see:
> • Simultaneous Equations:
>   `isaacphysics.org/concepts/cm_algebra_simult`

1. a) Find p and q if $q = 4p + 8$ and $3q + 6p = 6$.

    b) Find r and s if $3s + 4r = 7$ and $2s = 3r - 4$.

2. Find a and b if $a + b = 3$ and $2(a - 1) = b - 1$.

3. Find, in terms of m_1, m_2 and r, expressions for a) l_1 and b) l_2 when $l_1 + l_2 = r$ and $m_1 l_1 = m_2 l_2$. (This gives the distances from their centre of mass of two masses a distance r apart.)

4. Starting with the equations of motion $v = u + at$ and $s = \frac{1}{2}(u + v)t$, eliminate v to give the expression for s in terms of u, a and t.

5. Find a pair of values of a and b which solve $b = 3a^2 + 3$ and $b = 2a + 4$.

6. Find a pair of values of r and s which solve $r = 2s^2 - 3s + 3$ and $r = 2s + 1$.

7. Find p and q if $p^2 + 2pq + 4q^2 = 7$ and $2p = q + 1$.

8. Starting with the equations $v = u + at$ and $s = ut + \frac{1}{2}at^2$, eliminate t to find an equation relating s, u, v and a. Give your answer as an equation with v^2 only on the left hand side.

9. A particle of mass $2M$ is travelling at speed u towards a stationary particle of mass M and collides head-on and elastically with it. After the collision both particles are moving – the mass $2M$ with speed v and the mass M with speed w. Using the laws of conservation of momentum and kinetic energy we can write down two simultaneous equations for the collision: $2Mu = 2Mv + Mw$ and $\frac{1}{2}(2M)u^2 = \frac{1}{2}(2M)v^2 + \frac{1}{2}Mw^2$.

    a) Find an expression for w, in terms of u.

    b) Find the corresponding expression for v, in terms of u.

10. A particle of mass M, travelling at speed u, collides head-on and elastically with a stationary particle of mass m. After the collision the particles of mass M and of mass m travel at (non-zero) speeds v and w respectively.

By applying the laws of conservation of momentum and kinetic energy we can write down two simultaneous equations for the collision: $Mu = Mv + mw$ and $\frac{1}{2}Mu^2 = \frac{1}{2}Mv^2 + \frac{1}{2}mw^2$

a) Find an expression for w, the speed of the particle of mass m after the collision, in terms of u, M and m.

b) Find the corresponding expression for v, the speed of the particle of mass M after the collision, in terms of u, M and m.

1.4   Trigonometry – angles, triangles

In addition to the concepts from previous sections, see
- Geometry - Angles and Triangles - Level 1:
  `isaacphysics.org/concepts/cm_geometry1`

1. Fig. 1.1a shows a triangle of side lengths a, b and c with angles A, B and C.

   a) Find the angle A if $B = 30°$ and $C = 70°$.

   b) Find the angle D if $A = 40°$ and $B = 60°$.

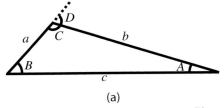

(a)

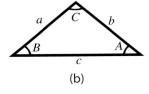

(b)

Figure 1.1

2. Fig. 1.1a shows a triangle of side lengths a, b and c with angles A, B and C.

   a) Find the angle A if $a = 10.0$ mm, $b = 14.0$ mm and $B = 65.0°$.

   b) Find the length c if $a = 10.0$ mm, $b = 6.00$ mm and $C = 40.0°$.

3. Fig. 1.1a shows a triangle of side lengths a, b and c with angles A, B and C.

   a) Find the length b if $a = 10.0$ mm, $A = 30.0°$ and $B = 70.0°$.

   b) Find the length c if $a = 10.0$ mm, $A = 60.0°$ and $B = 40.0°$.

4. Fig. 1.1a shows a triangle of side lengths a, b and c with angles A, B and C.
   Find the angle C if $a = 10.0$ mm, $b = 6.00$ mm and $c = 7.00$ mm.

5. Fig. 1.1a shows a triangle of side lengths a, b and c with angles A, B and C.

   a) Find the area of a triangle with $a = 10$ mm, $b = 4.0$ mm and $C = 70°$.

   b) Find the angle C if the area is 15.0 mm^2, $a = 10.0$ mm and $b = 4.00$ mm.

6. Fig. 1.1b shows an isosceles triangle of side lengths a, b and c, where $a = b = 10$ mm, and angles A, B and C, where $B = 50°$.

   a) Deduce the value of angle A.   b) Deduce the value of angle C.

   c) What is the side length c?

7. In Fig. 1.2a, OA and OB are radii of the circle centred at O, and the line EC is the tangent to the circle at B. Find the following angles.

   a) The angle OBA in terms of θ.   b) The angle ABD in terms of θ.

   c) The angle FBG in terms of θ.   d) The angle EBF in terms of θ.

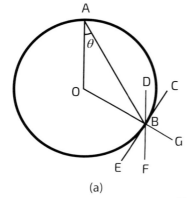

(a)

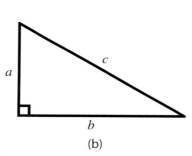

(b)

Figure 1.2

8. For a right-angled triangle, using the notation of Fig. 1.2b, find the length of the hypotenuse c in the following cases:

   a) when $a = 3$ cm, $b = 4$ cm,   b) when $a = 5.0$ cm, $b = 12$ cm,

   c) when $a = b = 3.00$ cm.

9. For a right-angled triangle, using the notation of Fig. 1.2b, find the length of the side a (to 2 sf) in the following cases:

   a) when $b = 6.0$ cm, $c = 10$ cm,   b) when $b = 10$ cm, $c = 26$ cm,

   c) when $b = 3.0$ cm, $c = 6.0$ cm.

10. An equilateral triangle has a perpendicular height of 2.00 cm.

   a) Find the length of the sides.   b) What is the area of the triangle?

1.5　Functions – evaluating, transforming, sketching

In addition to the concepts from previous sections, see:
- Functions - Simple Function Types - Level 1:
  isaacphysics.org/concepts/cm_functions_simple
- Functions - Transformations - Level 1:
  isaacphysics.org/concepts/cm_functions_transformations
- Trigonometric Relationships - Level 1:
  isaacphysics.org/concepts/cm_trig1

1. Consider the function $g(t) = (t-2)^3$. Find:

   a) $g(2)$,　　　b) $g(4)$,　　　c) $g(-1)$.

2. Consider the function $f(z) = 2z^2 - 4$. Find:

   a) $f(1)$,　　　b) $f(3)$,　　　c) $f(-3)$.

3. Consider the function $h(p) = \dfrac{2}{p} + 1$. Find:

   a) $h(1)$,　　　b) $h(-1)$,　　　c) $h(-2)$.

4. Consider the function $f(\theta) = 2\sin(2\theta)$. Find:

   a) $f(45°)$,　　　b) $f(-45°)$,　　　c) $f(90°)$.

5. Consider the function $k(\phi) = 3\tan(\phi + 90°)$. Find:

   a) $k(-90°)$,　　　b) $k(-45°)$,　　　c) $k(45°)$.

6. Sketch the graph of the function $g(x) = (x-2)^2 + 2$.

7. Sketch the graph of the function $f(x) = 3 - 2x$.

8. Sketch the graph of the function $f(x) = \dfrac{2}{x-1} + 1$.

9. Sketch the graph of the function $h(x) = \cos(2x) - 1$.

10. Sketch the graph of the function $k(x) = 2\sin(x - 45°) - 1$.

For more practice go to Section 2.5.

1.6 Graph Sketching – simple function types, trig functions

In addition to the concepts from previous sections, see
- Graph Interpreting - Level 1:
  `isaacphysics.org/concepts/cm_graph_interpreting`
- Graph Sketching - Level 1:
  `isaacphysics.org/concepts/cm_graph_sketching`

Toolkit #1. It is helpful to think about the following when interpreting a function $f(x)$.
1. What is the main shape of the function? (linear, quadratic, cubic, reciprocal, trigonometric, etc.)
2. Where are the zeros i.e. the values of x for which $f(x) = 0$?
3. What happens to the function when $x = 0$?
4. What happens to the function as x gets very large?
5. Does the function diverge anywhere within its range.

1. The graph of $y = f(x)$ is shown in Fig. 1.3a. Answer the **toolkit** questions about the function $f(x)$.

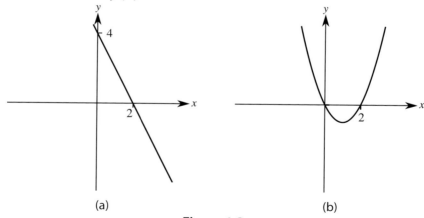

(a)   (b)

Figure 1.3

2. The graph of $y = f(x)$ is shown in Fig. 1.3b. Answer the **toolkit** questions about the function $f(x)$.

3. The graph of $y = f(x)$ is shown in Fig. 1.4a. Answer the **toolkit** questions about the function $f(x)$.

4. The graph of $y = f(x)$ is shown in Fig. 1.4b. Answer the **toolkit** questions about the function $f(x)$.

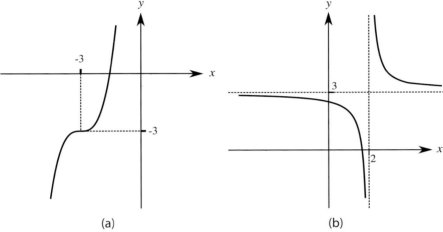

(a)   (b)

Figure 1.4

5. The graph of $y = f(x)$ is shown in Fig. 1.5. Answer the **toolkit** questions about the function $f(x)$.

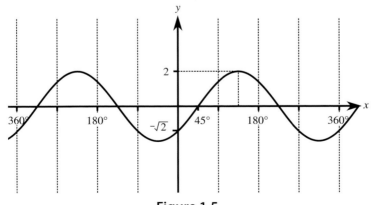

Figure 1.5

6. Answer the following questions about the linear functions $y = 2x + 3$ and $3y + 4x = 12$. Hence sketch graphs of the functions, labelling the key points on both axes.

   a) Consider $y = 2x + 3$. Deduce the value of x at which $y = 0$ (give your answer as an improper fraction) and the value of y when $x = 0$.

   b) Consider $3y + 4x = 12$. Deduce the value of x at which $y = 0$ and the value of y when $x = 0$.

7. Answer the following questions about the functions $f(x) = x^2 - 2x + 3$ and $g(x) = -2(x^3 + 1)$. Hence sketch the graphs of $y = f(x)$ and $y = g(x)$, labelling the key points on both axes.

   a) Deduce the value of $f(x)$ when $x = 0$.

   b) By completing the square deduce i. the value of x at which $f(x)$ has its minimum value and ii. the minimum value of $f(x)$.

   c) Deduce the value of $g(x)$ when $x = 0$.

   d) Deduce the value of x at which $g(x) = 0$.

8. Answer the following questions about the functions $f(x) = \dfrac{1}{x}$ and $g(x) = \dfrac{1}{1-x}$. Hence, using the same axes, sketch $y = f(x)$ and $y = g(x)$ between $x = -3$ and $x = 3$ and covering the range $y = -5$ to $y = 5$. Pay close attention to the shapes of the functions, label the key points on both axes and the point of intersection of the two functions.

   a) Answer the **toolkit** questions 2 to 5 about the function $f(x) = \dfrac{1}{x}$.

   b) Answer the **toolkit** questions 2 to 5 about the function $f(x) = \dfrac{1}{(1-x)}$.

   c) Find the x and y coordinates of the point of intersection of the functions $1/x$ and $1/(1 - x)$. Label this point on your sketch graph.

9. A mass on the end of a spring is oscillating up and down so that at time t its height y above the ground is given by $y = p + q \cos(rt)$, where y is in metres and t is in seconds, and $p = 2.0$ m, $q = 0.50$ m and $r = 90°\,\text{s}^{-1}$. Find the height above the ground at $t = 0$ s, $t = 1$ s, $t = 2$ s and $t = 3$ s. Hence, sketch and label a graph of the height of the mass above the ground as a function of time from $t = 0$ to $t = 8$ s.

10. The mass m of a sphere, as a function of its radius x, is given by the formula

$$m(x) = \frac{4}{3}\pi\rho x^3$$

where ρ is the density of the sphere.

A balance has a scale which has not been zeroed properly, so that a reading of -95 g is displayed when nothing is placed on the balance. A number of spheres, made of the same plastic but with varying radius, are weighed individually using this balance. The apparent mass $M(x)$ of these spheres with respect to their radius x is given by

$$M(x) = ax^3 + b$$

If M is measured in g and x in cm find a and b to 2 sf given that the density of the plastic is roughly 1200 kg m^{-3}. Hence, sketch a graph of M as a function of x in the range $x = 0$ to $x = 5$ cm. Pay close attention to the shape of the function and label the key points on both axes.

Level 2

In addition to the concepts from previous sections, see:
- Algebraic Manipulation - Inequalities - Level 2:
  isaacphysics.org/concepts/cm_algebra_manip_inequalities
- Algebraic Manipulation - Index Notation - Level 2:
  isaacphysics.org/concepts/cm_algebra_manip_index

1. Rationalise the denominators of the following expressions.

   a) $\dfrac{3\sqrt{6}}{2\sqrt{18}}$.

   b) $\dfrac{4-\sqrt{3}}{4+2\sqrt{3}}$.

2. Simplify the following expressions.

   a) $2\sqrt{20}+\sqrt{45}-5\sqrt{5}$.

   b) $4(\sqrt{3}+1)(\sqrt{3}-1)-2(2+\sqrt{2})(1+\sqrt{2})$.

3. Simplify the following expressions.

   a) $(4a^2b^3)^{\frac{1}{2}}\times(9ab^2)^{-\frac{3}{2}}$.

   b) $(8p^3q^2)^{\frac{2}{3}}\div(\dfrac{2p}{q^{\frac{1}{3}}})^5$.

   c) $(10^{-34})^{\frac{1}{2}}(10^{-10})^{\frac{1}{2}}(10^8)^{-\frac{5}{2}}$.

4. Solve the following inequalities.

   a) $3m+8\geq 2$.

   b) $2p+5<4p-7$.

5. Solve the following inequalities.

   a) $7-4a\leq -5$.

   b) $3-2(b+1)\geq 6+3(2b+1)$.

6. Solve the following inequalities.

   a) $3x^2-2x-8\leq 0$.

   b) $-2x^2+5<7x+11$.

7. A mass m is suspended on a spring with spring constant k in a medium which damps its motion. The condition that it will oscillate after it has been displaced from equilibrium is $\dfrac{k}{m} > \dfrac{b^2}{4m^2}$ where b is called the damping constant. Find the range of masses over which it will oscillate.

8. A body of mass m and speed v can escape from a planet of mass M and radius R if the sum of its kinetic energy ($\frac{1}{2}mv^2$) and its gravitational potential energy ($-GMm/R$) is greater than or equal to zero, i.e. $\frac{1}{2}mv^2 - \dfrac{GMm}{R} \geq 0$ (G is the universal constant of gravitation).

    a) Find the range of speeds v over which it will escape. Give your answer as an inequality, with v on the left hand side.

    b) If the speed of the body has a fixed value, i.e. $v = v_0$, and the mass of the planet $M = \frac{4}{3}\pi R^3 \rho$, where ρ is its average density, find the range of radii R for which the body will escape.

> The dimensions of physical properties do not depend on specific units; here we use length L, time T and mass M as our fundamental dimensions. In any equation relating physical properties the dimensions must be the same on both sides. For example $force = mass \times acceleration$. Obviously mass has dimensions M. To deduce the dimensions of acceleration recall that acceleration = change in velocity over time; velocity (= change in displacement over time) has dimensions of LT^{-1} so acceleration has dimensions $(LT^{-1})(T^{-1}) = LT^{-2}$. Thus force has dimensions MLT^{-2}.

9. a) The kinetic energy of a body of mass m moving with speed v is equal to $\frac{1}{2}mv^2$. Find the dimensions of (kinetic) energy. Recall that the factor of $\frac{1}{2}$ in the expression is dimensionless.

    b) One type of "Planck unit" is defined as $h^{\frac{1}{2}}G^{\frac{1}{2}}c^{-\frac{5}{2}}$, where h is Planck's constant (dimensions ML^2T^{-1}), G is the universal constant of gravitation (dimensions $M^{-1}L^3T^{-2}$) and c is the speed of light (dimensions LT^{-1}). Find the dimensions of this "Planck unit".

For more practice go to Section 5.7.

2.2   Trigonometry – sin, cos, tan, triangles

In addition to the concepts from previous sections, see:
- Trigonometric Relationships - Level 3:
  `isaacphysics.org/concepts/cm_trig1`

1. For the range $-180° \leqslant \alpha \leqslant 180°$, consider all the values of α which satisfy $\sin \alpha = 0.2$.

   a) How many values of α, satisfying the equation, are in this range?

   b) What is the largest positive value of α satisfying the equation in this range? Give your answer to 3 sf.

2. For the range $-180° \leqslant \beta \leqslant 180°$, consider all the values of β which satisfy $\sin(2\beta) = -0.4$.

   a) How many values of β, satisfying the equation, are in this range?

   b) What is the smallest positive value of β in this range? Give your answer to 3 sf.

3. For the range $-180° \leqslant \alpha \leqslant 180°$, consider all the values of α which satisfy $\cos \alpha = -0.7$.

   a) How many values of α, satisfying the equation, are in this range?

   b) What is the largest value of α satisfying the equation in this range? Give your answer to 4 sf.

4. For the range $-180° \leqslant \alpha \leqslant 180°$, consider the values of α which satisfy the equation $\cos 4\alpha = 1$.

   a) How many values of α, satisfying the equation, are in this range?

   b) What is the smallest (in magnitude) negative value of α satisfying the equation in this range?

5. For the range $0 \leqslant \theta \leqslant 360°$, write down all the values of θ which have the following: a) $\sin \theta = \sqrt{3}/2$, b) $\sin \theta = -1/2$.

6. For the range $0 \leqslant \theta \leqslant 360°$, write down all the values of θ which have the following: a) $\cos \theta = -1/2$, b) $\cos \theta = 1/\sqrt{2}$.

7. Write down the exact values of the following: a) $\cos 150°$, b) $\cos 300°$.

8. Write down the exact values of the following: a) $\sin 120°$, b) $\sin 225°$.

9. Write down the exact values of the following: a) $\tan 135°$, b) $\tan 240°$.

10. Figure 2.1 shows a right-angled triangle with hypotenuse OP.

   a) Find OP if PQ $= 4.00$ cm and $A = 35.0°$.
   b) Find PQ if OQ $= 4.00$ cm and $A = 40.0°$.
   c) Finally, find OQ if OP $= 4.00$ cm and $A = 50.0°$.

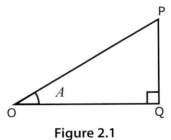

Figure 2.1

2.3   Simple Shapes – area, volume

In addition to the concepts from previous sections, see:
- Geometry - Shapes and their Properties - Level 2:
  `isaacphysics.org/concepts/cm_geometry2`

1. A cylinder of radius R contains some liquid; a solid metal cuboid with sides a, a and $2a$ is totally immersed in the liquid. Find an expression for the decrease in the height of the liquid in the cylinder when the cuboid is taken out of the liquid.

2. A cylinder of radius $2a$ contains water to a depth h. A cone with base radius a and perpendicular height a is lowered into the water. When the bottom of the cone is resting on the bottom of the beaker the water just covers the top of the cone. Find an expression for the initial depth of the water h in terms of a.

3. Four identical spherical water droplets coalesce to form one spherical drop. Find the factor (to 3 sf) by which the total surface area changes.

4. Find the volume of a sphere of radius 2.00 cm in:

   a) cm^3,   b) mm^3,   c) m^3.

5. Find the volume of a sphere which has a surface area of 0.45 m^2.

6. A sphere has volume V.

   a) Find an expression for the radius r of the sphere in terms of V.

   b) What is the surface area A in terms of V?

7. A cylinder with closed ends has a total surface area S; the radius of the base is a and the height is ka. Find an expression for a in terms of k and S.

8. A hollow steel cylinder of length 8.0 cm is to be constructed with an inner radius of 1.9 cm and an outer radius of 2.0 cm, using steel of density 7800 kg m^{-3}.

   a) Find the required volume of steel.

   b) Find the required mass of steel.

9. The pyramid in Fig. 2.2a has a square base of side a. The four identical tri-angles making up its sides each have a base of length a and perpendicular height b.

a) Find in terms of a and b the total surface area S of the pyramid.

b) Find in terms of a and b the total volume V of the pyramid.

c) A cone has the same volume and height h as the pyramid; find, in terms of a, the radius r of its base.

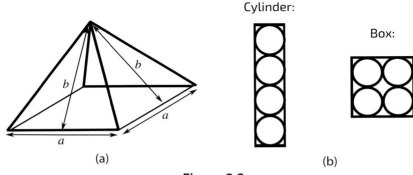

Cylinder:

Box:

(a)   (b)

Figure 2.2

10. a) Four tennis balls are placed one above the other so that they just fit into a cylindrical tube as shown in Fig. 2.2b (i.e. each ball is in contact with the sides of tube and the top and the bottom balls are in contact with the lid and the base of the cylinder respectively). Find the fraction of the space inside the cylinder occupied by the balls. Give your answer to 3 sf.

b) The same four balls are now placed in a box with a square cross-section so that they are touching the sides of the box and each other as shown in Fig. 2.2b and are in contact with the lid and the base of the box. Find the fraction of the space inside the box occupied by the balls. Give your answer to 3 sf.

2.4   Vectors – notation; adding, resolving components

In addition to the concepts from previous sections, see:
- Vectors - Resolving Vectors - Level 2:
  isaacphysics.org/concepts/cm_vectors2
- Vectors - Describing and Adding Vectors - Level 2:
  isaacphysics.org/concepts/cm_vectors
- Potential Energy - Level 1: Constant g Field:
  isaacphysics.org/concepts/cp_potential_energy

1. Two vectors p and q are given by $p = \begin{pmatrix} 4 \\ 3 \end{pmatrix}$ and $q = \begin{pmatrix} 2 \\ -1 \end{pmatrix}$. Find the following to 3 sf.

   a) The magnitude of p.

   b) The magnitude of q.

   c) The angle that p makes with the x-axis.

   d) The vector $p + q$ in column vector form and give its magnitude.

   e) The column vector form of $p - q$ and the angle this vector makes with the x-axis.

2. The vector $u = \begin{pmatrix} -3 \\ 6 \end{pmatrix}$. Find an expression using unit vector notation for the unit vector $\hat{u}$ in the direction of u.

3. A vector a is of length 8.00 and makes an angle of $40°$ to the x-axis. What is the magnitude of its projection onto the x-axis and the y-axis? Now, express a using unit vector notation.

4. A boat sails 4.00 km at a bearing of $210°$.

   a) How far south of its starting point is its final position?

   b) How far west of its starting point is its final position?

5. A block of weight W is on a slope which makes an angle α to the horizontal as shown in Fig. 2.3a.

   a) What is the component of the block's weight W that acts perpendicular to the slope?

   b) The block slides a distance l along the slope. How far has the block moved horizontally?

   c) The block slides a distance l along the slope as in part b). How much gravitational potential energy has the block lost?

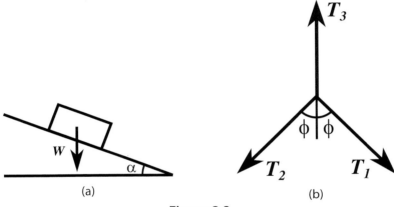

(a)   (b)

Figure 2.3

6. A body is acted on by three forces T_1, T_2 and T_3 in the (x, y)-plane as shown in Fig. 2.3b, where $T_3 = 20.0\hat{\jmath}$ N, $\phi = 20.0°$ and the body is equilibrium.

   a) Find the magnitude of T_1.

   b) Find the x component of T_2, taking left-to-right to be the positive direction.

7. The coordinates A, B and C of the vertices of a triangle are given by $(1, 1, -1)$, $(-1, 1, 1)$ and $(1, -1, 1)$ respectively. Find the following.

   a) The vector from A to B, expressed in unit vector notation.

   b) The vector from B to C, expressed in unit vector notation.

   c) The vector from C to A, expressed in unit vector notation.

   d) The side length (to 3 sf) of the equilateral triangle ABC.

8. Find the vector going from the point $(2,2,0)$ to $(1,4,3)$.

9. The coordinates of one vertex A of a triangle ABC are $(3,4,2)$ and the vectors from A to the other vertices B and C are $(1,1,4)$ and $(-3,2,-2)$ respectively.

   a) Find the coordinates of B.

   b) Find the coordinates of C.

   c) Find the length (to 3 sf) of the side a, i.e. the vector from B to C.

10. A vector $\begin{pmatrix} u_x \\ u_y \\ u_z \end{pmatrix}$ has a length of 4.00 units. Answer the following questions.

   a) The vector lies in the (x,y)-plane, makes an angle of $30°$ with the x-direction and u_y is positive. What is u_x?

   b) The vector has $u_x = u_y = 2.00$ and u_z is negative. What is u_z?

   c) The vector is such that $u_z = 1.00$ and $u_y = 2u_x$. What is u_y?

For more practice go to Section 5.1.

2.5   Functions – polynomials, symmetry; transforming

In addition to the concepts from previous sections, see:
- Functions - Polynomials and Rational Functions - Level 2:
  `isaacphysics.org/concepts/cm_functions_polynomials`

1. The electrostatic potential $V(x,y)$ at a point in the (x,y)-plane with co-ordinates (x,y) due to a point charge lying in the (x,y)-plane at the origin, i.e. with coordinates $(0,0)$, is given by $V(x,y) = \dfrac{P}{\sqrt{x^2+y^2}}$ where P is a constant proportional to the magnitude of the charge. Find expressions for:

   a) $V(a,0)$,   b) $V(0,b)$,   c) $V(a,b)$,   d) $V(a,a)$.

2. Investigate the transformations of the following functions.

   a) The functions $f(x) = x^2 + 2x + 1$ and $g(x) = f(x-a)$, where a is a constant. If $g(1) = 9$, find the value of a, given that it is positive.

   b) The functions $r(u) = \dfrac{2}{u-2}$ and $s(u) = r(u) + b$, where b is a constant. If $s(0) = 1$, find the value of b.

   c) The functions $p(r) = \dfrac{1}{r}$ and $q(r) = p(r-c) + d$, where c and d are constants. If $q(0) = 1$ and $q(2) = 3$, find the values of c and d.

3. A particle of mass M moving at speed u collides head on with a stationary particle of mass m. No energy is lost in the collision. The speeds after the collision depend upon the relative magnitudes of M and m. Taking M as fixed we can express these speeds as a function of m. The speed $v(m)$ of mass M after the collision is given by $v(m) = \dfrac{M-m}{M+m}u$, and the speed $w(m)$ of mass m is given by $w(m) = \dfrac{2M}{M+m}u$. Find the expressions for v and w in terms of u when:

   a) $m = M$,   b) $m = 2M$,
   c) $m = M/2$,   d) $m = rM$ where r is any positive number.

4. Investigate the transformations of the following functions.

a) $f(x) = x^2 + 2x - 1$ is transformed into another function $g(x)$ by stretching it by a factor a in the x-direction, i.e. $g(x) = f(\frac{x}{a})$. If $g(2) = 2$, what is the value of a? (You may assume that a is positive).

b) $v(u) = \dfrac{3}{1 + 2u}$ is transformed into $w(u)$ by stretching it by a factor b in the vertical direction, i.e. $w(u) = bv(u)$. If $w(4) = \frac{1}{2}$, find b.

c) $f(x) = 2x + 3$ is transformed into $g(x)$ by stretching it by a factor p in the lateral direction and then by a factor of q in the vertical one, i.e. $g(x) = qf\left(\dfrac{x}{p}\right)$. If $g(1) = 2$ and $g(-2) = -3$, find the equation for $g(x)$.

The following question asks you to deduce the symmetry properties of a number of functions. There are three choices: 1) even - a function for which $f(x) = f(-x)$ which is also described as being symmetric about the vertical axis, 2) odd - a function for which $f(x) = -f(-x)$ which is also described as being antisymmetric about the vertical axis (or symmetric about zero), 3) neither even nor odd.

---

5. Decide the symmetry properties of the following functions. Where relevant you may assume that a and b are non-zero constants.

a) Decide which of the following functions are even:

ax^2, $ax^2 + b$, $ax^2 + bx^4$, $\dfrac{a}{x^2} + bx^2$, $(x-a)(x+a)$, $a\cos x$, $a(x+b)^2$, $\dfrac{a}{x^2} + b$, $(x-a)(x+b)$ $(a \neq b)$, $a\sin x$, $x^2(a+bx)$.

b) Decide which of the following functions are odd:

ax, $\dfrac{a}{x}$, $\dfrac{a}{x} + bx^3$, $x(a+bx^2)$, $x^{1/3}$, $a\sin x$, $x^2(a+bx)$, $(x+a)^{1/3}$, $a\tan x$, $\dfrac{a}{x} + b$, $\dfrac{a}{x} + \dfrac{b}{x^3}$.

c) Decide which of the following functions are neither odd nor even:

$ax - b$, $x^2(ax+b)$, $(x-a)(x+a)^2$, $\dfrac{a}{(x-b)^2}$, $ax^{1/2}$, $a\tan(x + 45°)$, $a\left(\dfrac{1}{x^2} - \dfrac{1}{b^2}\right)$, $a(b-x)^{1/2}$, $\cos x + \sin x$, $x(ax^2+b)$, $(x-a)(x+a)$.

6. A sinusoidal wave is travelling in the positive x-direction. Its displacement $\psi(x,t)$ at a point x at time t is given by $\psi(x,t) = \psi_0 \cos(\frac{2\pi x}{\lambda} - 2\pi ft)$ where ψ_0 is the amplitude, λ is the wavelength and f is the frequency of the wave; the speed of the wave $v = f\lambda$. Find expressions in terms of ψ_0, λ, f, x and t (as appropriate) for:

a) $\psi(0,0)$,   b) $\psi(x + vt, t)$,   c) $\psi(x, t + \frac{x}{v})$.

7. Consider the polynomials $q(p) = p^4 + 2p^2 + 3p + 2$ and $r(p) = p^4 - 3p^3 - 2p^2 + 3$. Find:

a) $q(p) + r(p)$,   b) $q(p) - r(p)$.

8. Consider $f(t) = t^3 - 2t^2 + 3t - 1$ and $g(t) = 4t^2 + 2t + 2$. Express each of the following as a single polynomial:

a) $2f(t) + g(t)$,   b) $2f(t) - 3g(t)$.

9. Find the following polynomial products:

a) $(s^2 - 1)(s^2 - 2s + 3)$,   b) $(u - 1)(3u - 2)(2u + 3)$.

10. Consider the polynomial product $(au^2 + bu + c)(du^3 + eu^2 + fu + g)$ where a, b, c, d, e, f and g are constants. Find expressions in terms of these constants for the coefficients of u^3 and u.

For more practice go to Section 3.2.

2.6 Differentiation – powers, stationary points

In addition to the concepts from previous sections, see:
- Calculus - Differentiation - Level 2:
  `isaacphysics.org/concepts/cm_differentiation`

1. a) Find $\dfrac{dy}{dx}$ if $y = x^4$.

   b) Find the gradient of the curve $x = t^2$ at the points $t = 0$, $t = 3$ and $t = -3$.

2. a) Find $\dfrac{dv}{du}$ if $v = \dfrac{1}{u}$.

   b) Find the equation of the tangent to the curve $v = \dfrac{1}{u}$ at $u = 2$.

   c) Find $\dfrac{dF}{dr}$ if $F = Ar^3$, where A is a constant.

3. a) Find the gradient of $w = z^{1/2} + z^{-1/2}$ at $z = 1/4$, $z = 1$ and $z = 4$.

   b) Find the roots of the curve $f(u) = u^3 - 4u$.

   c) Find the gradient at each of the roots of $f(u)$.

4. a) Find the gradient of the curve $t = 4s^{-3/4}$ at the point $s = 16$.

   b) Find $\dfrac{dx}{dt}$ if $x = bt^{3/2}$.

   c) Find $\dfrac{d^2x}{dt^2}$ if $x = bt^{3/2}$.

5. a) Differentiate $ax^3 + (b/x) + c$ with respect to x, where a, b and c are constants.

   b) Differentiate $(2m+3)(m-1)$ with respect to m.

6. a) Find $\dfrac{dv}{du}$ if $v = Bu^{-3}$.

   b) The electrostatic potential energy V of two equal charges q a distance r apart is given by $V = \dfrac{q^2}{(4\pi\epsilon_0 r)}$. The force between the two charges is given by $-\dfrac{dV}{dr}$; find an expression for this force. Interpret the sign.

7. Find the coordinates, nature and number of the stationary points of the following functions.

a) $y = 2x^3 - 24x - 5$.   b) $y = 2x^3 - 5x^2 + 4x + 6$.

8. Find the coordinates, nature and number of the stationary points of the following functions.

a) $p = q + (4/q)$.

b) $t = ay^4 - by^2 + c$, where a, b and c are all positive.

9. a) A particle is moving in one dimension. Its displacement s at time t is given by $s = ut + bt^2$. The velocity v of the particle at time t is given by the rate of change of displacement with time, i.e. $v = \dfrac{ds}{dt}$. Find an expression for the velocity.

b) The acceleration a of the particle at time t is given by the rate of change of velocity with time. Find an expression for the acceleration of the particle in part a).

c) The displacement of a body at time t is given by $x = \alpha t + \beta t^3$ where $\alpha = 4$ m s^{-1} and $\beta = 5$ m s^{-3}. Find the velocity of the body at $t = 2$ s.

d) Find the acceleration of the body in part c) at $t = 2$ s.

10. a) A particle is fired upwards into the air with a speed w and moves subsequently under the influence of gravity with an acceleration g downwards, such that its height h at time t is given by $h = wt - \frac{1}{2}gt^2$. Find an expression for its maximum height above its initial position.

b) The potential energy of two molecules separated by a distance r is given by $U = U_0 \left(\left(\dfrac{a}{r}\right)^{12} - 2\left(\dfrac{a}{r}\right)^6 \right)$ where U_0 and a are positive constants. The equilibrium separation of the two molecules occurs when the potential energy is a minimum; find expressions for the equilibrium separation and the value of the potential energy at this separation.

For more practice go to Section 3.4.

2.7 Graph Sketching – powers of x, polynomials

In addition to the concepts from previous sections, see
- Graph Interpreting - Level 2:
  isaacphysics.org/concepts/cm_graph_interpreting
- Graph Sketching - Level 2:
  isaacphysics.org/concepts/cm_graph_sketching

Toolkit #2. It is helpful to think about the following when interpreting a function $f(x)$.
1. What is the main shape of the function? (linear, quadratic, cubic, reciprocal, trigonometric etc.)
2. Where are the zeros i.e. the values of x for which $f(x) = 0$?
3. What happens to the function when $x = 0$?
4. What happens to the function as x gets very large?
5. Does the function diverge anywhere within its range.
6. What are the symmetry properties of the function?

1. Fig. 2.4a shows a sketch of the variation of the gravitational field $g(r)$ with distance r from the centre of a spherical planet of radius R. The field varies with r in a different way inside and outside the planet. Identify the correct functional form of this field from the choices below; A is a constant.

$$g(r) = \begin{cases} Ar & \text{for} \quad r \leq R \\ \dfrac{AR^3}{r^2} & \text{for} \quad r > R \end{cases} \qquad g(r) = \begin{cases} \dfrac{AR^3}{r^2} & \text{for} \quad r \leq R \\ Ar & \text{for} \quad r > R \end{cases}$$

$$g(r) = \begin{cases} Ar & \text{for} \quad r \leq R \\ \dfrac{AR^2}{r} & \text{for} \quad r > R \end{cases} \qquad g(r) = \begin{cases} Ar & \text{for} \quad r \leq R \\ \dfrac{Ar^2}{R^3} & \text{for} \quad r > R \end{cases}$$

$$g(r) = \begin{cases} \dfrac{A}{r} & \text{for} \quad r \leq R \\ Ar & \text{for} \quad r > R \end{cases} \qquad g(r) = \begin{cases} Ar & \text{for} \quad r \leq R \\ \dfrac{A}{r^2} & \text{for} \quad r > R \end{cases}$$

If you are not familiar with the notation used in these answers,

$$g(r) = \begin{cases} Ar^n & \text{for} \quad r \leq R \\ \dfrac{AR^m}{r^2} & \text{for} \quad r > R \end{cases}$$

means that $g(r) = Ar^n$ for $r \leq R$ and $g(r) = \dfrac{AR^m}{r^2}$ for $r > R$, where n and m are particular powers.

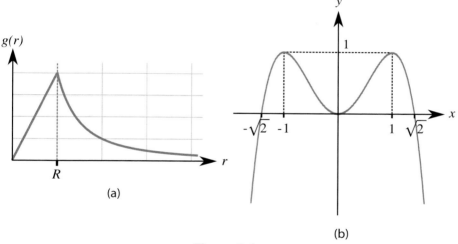

(a)

(b)

Figure 2.4

2. The graph shown in 2.4b is a polynomial of order 4, i.e. it has the form $f(x) = ax^4 + bx^3 + cx^2 + dx + e$ where a, b, c, d and e are constants. Use the information shown in the graph, including its symmetry, to deduce the values of the constants.

a) From the symmetry of the graph deduce the value of b and d.

b) Deduce the value of the constants e, a and c.

3. The magnetic field $B(r)$ at a perpendicular distance r from the centre of a current-carrying conductor of radius a is given by

$$B(r) = \begin{cases} \dfrac{Ar}{a^2} & \text{for} \quad r < a \\[2mm] \dfrac{A}{r} & \text{for} \quad r \geq a \end{cases}$$

Notice the two different functional forms inside and outside the wire. (If you are not familiar with the above notation it means that $B(r) = \dfrac{Ar}{a^2}$ for $r < a$ and $B(r) = \dfrac{A}{r}$ for $r \geq a$.) Which of the graphs in Fig. 2.5 represents this magnetic field?

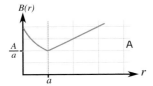 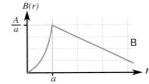 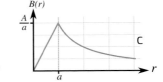

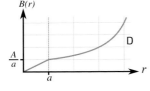 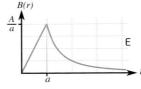 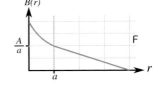

Figure 2.5

4. Answer the questions below about the cubic function $h(y) = y^3 - 8y^2 + 19y - 12 = (y-1)(y-3)(y-4)$. Hence, for a suitable range of values of y, sketch a graph of the function; consider its shape carefully and label the key points on both axes. (You are not expected to find the values of y and $h(y)$ at which the turning points of the function occur.)

a) Deduce the smallest value of y at which the function crosses the horizontal axis (i.e. where $h(y) = 0$).

b) Deduce the largest value of y for which $h(y) = 0$.

c) Deduce the value of h when $y = 0$.

d) What happens to the function when y gets very large?

5. Graphs of six functions are shown in 2.6. Deduce the symmetry properties of each function A to F, i.e. is it symmetric about the vertical axis (even), anti-symmetric about the vertical axis (odd, or symmetric about zero) or neither.

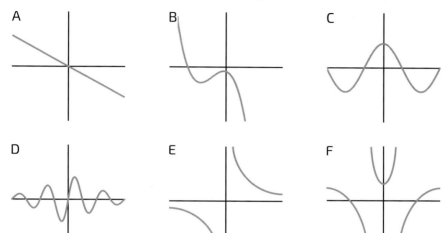

A   B   C

D   E   F

Figure 2.6

6. Answer the following questions about the functions $f(x) = x^{\frac{1}{2}}$ and $g(x) = x^{\frac{1}{3}}$. Hence using the same axes, sketch graphs of $y = f(x)$ and $y = g(x)$ over the range $x = -9$ and $x = 9$. Pay close attention to the differences in the shapes of the two functions, and label the key points on both axes and the point of intersection of the two functions.

a) Deduce the value of $f(x)$ when $x = 0$.

b) What are the symmetry properties of the function $f(x)$?

c) Deduce the value of $g(x)$ when $x = 0$.

d) What are the symmetry properties of the function $g(x)$?

e) Find the x coordinate of the point of intersection of the functions $x^{\frac{1}{2}}$ and $x^{\frac{1}{3}}$. Deduce the corresponding value of y. Label this point on your graph.

7. Consider the functions $f(x) = \dfrac{2}{x}$ and $g(x) = \dfrac{4}{x^2} - 2$.

a) Answer the **toolkit** questions about the function $f(x)$.

b) Answer the **toolkit** questions about the function $g(x)$.

c) Using the same axes, sketch the graphs of $y = f(x)$ and $y = g(x)$ in the range $x = -4$ to $x = 4$. Pay close attention to the differences in the shapes of the two functions, and label the key points on both axes and the points of intersection of the two functions.

8. A polynomial of order 4 has the form $f(x) = ax^4 + bx^3 + cx^2 + dx + e = a(x - p)(x - q)(x - r)(x - s)$ where a, b, c, d and e are constants, and $a < 0$. The constants p, q, r and s are such that $p < 0$ and $0 < q < r < s$. Answer the questions below about this function. Hence sketch a graph of the function; consider its shape carefully and label the key points on both axes. (You are not expected to find the values of x and $f(x)$ at which the turning points of the function occur.)

a) Give one of the values of x at which the function crosses the horizontal axis (i.e. where $f(x) = 0$)

b) Write down the relationship between e and a, p, q, r and s. From this you can deduce whether the value of f when $x = 0$ is positive or negative.

c) What happens to the function $f(x)$ when x gets very large?

9. A mass m is suspended on the end of a spring of spring constant k. The energy $U(x)$ stored in the system is made up of the elastic potential energy stored in the spring and the gravitational potential energy of the mass. It is measured relative to the point where the string is unstretched and is given by $U(x) = \frac{1}{2}kx^2 - mgx$, where x is the extension of the spring, measured downwards, and g is the acceleration due to gravity.

a) Find the values of x for which $U = 0$.

b) By considering the results from part a), or by completing the square, find the value of x at which the stored energy $U(x)$ is a minimum and deduce the value of U at that point.

c) Sketch the graph of $U(x)$ as a function of x; consider the shape of your function carefully and label the key points on both axes.

10. A pendulum consists of a small mass attached to one end of a light rod; the other end is attached to a pivot. The mass oscillates backwards and forwards and at time t its displacement from its equilibrium point vertically below the pivot is $x(t)$ and its speed is $v(t)$. At time t the displacement $x(t)$ of the mass from its equilibrium position is given by $x(t) = a\sin(\omega t + \phi)$, where x is in cm and t is in s, and $a = 3.0$ cm, $\omega = 90\,°\,s^{-1}$ and $\phi = 30\,°$. The speed $v(t)$ of the mass is given by $v(t) = u\cos(\omega t + \phi)$, where v is in cm s^{-1} and t is in s, and $u = 4.7$ cm s^{-1}, and $\phi = -30\,°$. Answer the questions below about $x(t)$ and $v(t)$. Hence sketch and label a graph of the displacement $x(t)$ and the speed $v(t)$ as a function of time from $t = -4$ s to $t = 4$ s.

a) Find the first time after $t = 0$ that the mass passes through its equilibrium position (i.e. $x(t) = 0$). (Give your answer to 2 sf.)

b) Find the second time after $t = 0$ that the mass passes through its equilibrium position. (Give your answer to 3 sf.)

c) Find the first time after $t = 0$ that the mass has its largest displacement from equilibrium (give your answer to 3 sf.) and deduce the maximum displacement of the mass.

d) Find the first time after $t = 0$ that the speed of the mass is zero (i.e. $v(t) = 0$). (Give your answer to 3 sf.)

e) Find the second time after $t = 0$ that the speed of the mass is zero. (Give your answer to 3 sf.)

f) Find the first time after $t = 0$ that the mass has its largest speed (give your answer to 2 sf.) and deduce the maximum speed of the mass.

Level 3

3.1   Trigonometry – circles, radians/degrees

In addition to the concepts from previous sections, see:
- Geometry - Shapes and their Properties - Level 3:
  `isaacphysics.org/concepts/cm_geometry2`
- Geometry - Angles and Triangles - Level 3:
  `isaacphysics.org/concepts/cm_geometry1`

1. a) For a circle of radius a, write down an expression for the length of the arc subtending an angle of ϕ (in radians) at the centre of the circle.

    b) For a circle of radius a, write down an expression for the angle (in radians) subtended at the centre of the circle by an arc of length b.

    c) The arc AB of a circle is of length s and the area of the sector AOB is A as shown in Fig. 3.1a; find an expression for the radius of the circle and the angle AOB in terms of s and A.

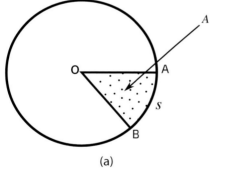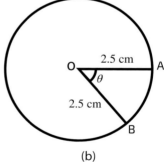

(a)   (b)

Figure 3.1

2. a) Figure 3.1b shows a circle with a radius of 2.5 cm. Find the length of the arc AB if $\theta = \frac{\pi}{3}$. Find the area of the sector AOB if $\theta = 80°$.

    b) Let the arc AB of the circle now have radius 3.0 cm. Find the angle AOB, the area of the sector AOB and the length of the chord AB.

3. Expressing your answer in terms of π, write the following angles in radians.

    a) $45°$ b) $60°$ c) $180°$ d) $240°$

4. Convert the following angles to radians.

   a) $0.100°$ b) $2.00°$ c) $40.0°$ d) $160°$

5. Give the following angles in degrees.

   a) $\frac{\pi}{2}$ rad   b) $\frac{2\pi}{3}$ rad   c) $\frac{3\pi}{4}$ rad   d) $\frac{7\pi}{6}$ rad

6. Convert the following to degrees. Give your answers to 3 sf.

   a) 0.0500 rad   b) 0.200 rad   c) 0.900 rad   d) 2.00 rad

7. Simplify the following trigonometric expressions to give a single trigonometric function.

   a) $\dfrac{1}{\cos^2(t) - 1}$.   b) $\dfrac{1 - \sin^2(x)}{\cos(x)}$.

   c) $\sin(\alpha)\tan(\alpha) - \dfrac{1}{\cos(\alpha)}$.   d) $\tan(w) - \dfrac{\cos(w)}{1 - \sin(w)}$.

8. Forces $F_1 = (F_{1x}, F_{1y})$ and $F_2 = (F_{2x}, F_{2y})$ act in the x-y plane, having magnitudes F_1 and F_2 and making angles of θ and ϕ with the positive x-axis
   respectively. Find expressions for the following.

   a) F_{1x} in terms of F_1 and θ.   b) F_{1y} in terms of F_1 and θ.

   c) F_{2x} in terms of F_2 and ϕ.   d) F_{2y} in terms of F_2 and ϕ.

   e) The magnitude of the vector sum of the two forces.

   f) The angle the vector sum of the two forces makes to the x-axis.

9. Two forces act on a body. One force of 20.0 N acts in a horizontal direction;
   the other force of 30.0 N acts in the vertical direction.

   a) Find the magnitude of the resultant force.

   b) What angle does the resultant force make to the horizontal?

10. a) Find, without using a calculator, $\cos(\phi)$ and $\tan(\phi)$ given that ϕ is an ob-
    tuse angle and that $\sin(\phi) = \dfrac{5}{13}$.

    b) A particle is projected into the air with a speed of 50 m s^{-1} at an angle of
    θ to the horizontal. It lands a horizontal distance of 250 m away after 6.4 s.
    Assuming that it travels at a constant velocity in the horizontal direction find
    the value of θ.

3.2   Functions – exponentials, logarithms

In addition to the concepts from previous sections, see:
- Functions - Exponentials and Logarithms - Level 3: `isaacphysics.org/concepts/cm_functions_exponentials_logarithms`

1. Solve the following for y : $4^y = 8^{y+1}$.

2. Solve the following for x : $3^x = \dfrac{1}{\left(9^{x-\frac{9}{4}}\right)}$.

3. Solve the following for m : $\dfrac{1}{9^m} = 27^{1-m}$.

4. a) Find the value of a for which $\log_a 4 = 2$.
   b) Find the value of a for which $\log_3 a = 3$.

5. Express the following in terms of $\log(u)$, $\log(v)$ and $\log(w)$ as required.
   a) $\log\left(\sqrt{\dfrac{uv}{w}}\right)$.   b) $\log(0.1u^2\sqrt{w})$.

6. Solve the following logarithmic equations.
   a) b if $\log_3 \sqrt{b} = 2$.   b) x if $\log_2(x^2) - \log_2 3 = \log_2 48$.

7. a) Express $\log 3 + \log 12 - \log 4$ in terms of $\log 3$.
   b) Express $\log 9 + 3\log 2 - 2\log 6$ in terms of a single logarithm.

8. Find:
   a) $\log \sqrt{0.0001}$,   b) $\log_2 \sqrt{1/2}$.

9. The apparent magnitude m of an astronomical object describes on a logarithmic scale how bright an object appears to an observer. It is related to its actual brightness or energy flux F (i.e. the energy arriving at the Earth per unit area per second) in the following way. Consider two objects with magnitudes m_1 and m_2 and brightnesses F_1 and F_2; the relationship between these quantities is $\frac{F_1}{F_2} = 100^{(m_2-m_1)/5}$.

a) The magnitude of the Sun is -26.8 and it is a factor of 4.8×10^5 brighter than the full Moon. Find the magnitude of the full Moon.

b) Supernova 1987A was discovered in the nearby dwarf galaxy the Large Magellanic Cloud and, with a magnitude of $+2.9$, it was visible with the naked eye. It was subsequently discovered that its progenitor was a blue supergiant with a magnitude of $+12.2$. Find the ratio of the brightness of Supernova 1987A to that of its progenitor to 2 sf.

10. A steel bar is tapped on one end and the resulting pulse of energy travels backwards and forwards along the bar. A very small fraction α of its energy is lost on each reflection so that after n reflections the fraction of its initial energy left is $(1 - \alpha)^n$. It takes a time τ to travel from one end of the bar to the other.

a) Find an expression for the time it takes for the energy in the pulse to halve.

b) Find an expression for the time it takes for the energy in the pulse to fall by a factor of 100.

For more practice go to Section 4.2.

3.3   Series – binomial expansion

In addition to the concepts from previous sections, see:
- Series - Level 3: `isaacphysics.org/concepts/cm_series`

1. Expand and simplify:

   a) $(x + 1)^4$.   b) $(z + 2a)^3$.   c) $(a - b)^5$.

2. Find the coefficient of x^3 in the expansion of:

   a) $(x - 10)^5$.   b) $(2x - \frac{1}{2})^6$.   c) $(x - y)^{10}$.   d) $(x - \frac{1}{x})^7$.

3. Without expanding the binomials, find:

   a) The coefficient of $x^4 y^6$ in the expansion of $(x^2 + 3y^2)^5$.

   b) The coefficient of x^{20} in the expansion of $(x^2 + 3x)^{12}$.

   c) The coefficient of ab^7 in the expansion of $(a + \frac{1}{4}b)^8$.

   d) The constant term in the expansion of $(\frac{x^2}{2} - \frac{8}{x})^9$.

4. Expand and simplify $(2 + x)^6$ in ascending powers of x up to and including x^3.

5. Expand $(2 + y)^5$ in ascending powers of y up to and including the term in y^3. Hence, without using a calculator, evaluate $(2.1)^5$ correct to 3 sf.

6. Expand $(3 - a)^4$ in ascending powers of a up to and including the term in a^3. Hence, without using a calculator, evaluate $(2.9)^4$ correct to 2 decimal places.

7. Expand $(1 - z)^3$. Hence, without using a calculator, evaluate $(0.98)^3$ to 3 decimal places.

8. Expand $(1 - 2x + 3x^2)^7$ in ascending powers of x as far as x^3.

9. What is the ratio of the $(r + 1)$th term to the rth term in the expansion of $(1 + \frac{x}{2})^n$?

10. Consider the expansion of $(1 + x)^n$.

    a) Find the sum of all coefficients.

    b) Show that the sum of the coefficients of terms to an odd power of x equals that of terms to an even power of x. Find the value of this sum.

3.4   Differentiation – powers, stationary points

1. A rectangular cuboid has a base with sides of length a and b and a height c. Its volume V and height c are fixed. By following the steps below, find expressions in terms of V and c for the values of a and b which will minimise the surface area A of the cuboid, find an expression for this minimum surface area and check that this is indeed a minimum.

a) Write down the equation for the volume V of the rectangular cuboid in terms of a, b and c. Then, write down the equation for the area A of the rectangular cuboid in terms of a, b and c. Now, from your equation for V deduce an expression for b in terms of V, a and c. Hence, by substitution, obtain an equation for A in terms of V, a and c.

b) Differentiate with respect to a the expression for A you found in part a) (since V and c are fixed you may treat them as constants). Hence find, in terms of V and c, an expression for a for which the area A is minimised.

c) Find, in terms of V and c, the expression for b for this value of a.

d) Find an expression for the minimum area in terms of V and c.

e) Find, at the value of a deduced in part b), an expression in terms of V and c for the second derivative of A with respect to a; convince yourself that the value of the second derivative indicates that A is a minimum at this point.

2. A disk of radius R is oscillating under gravity about an axis parallel to, and at a distance a from, the axis of the disk. The period of oscillation of the disk is given by $T = 2\pi \sqrt{\dfrac{a^2 + \frac{1}{2}R^2}{ga}}$ where g is the acceleration due to gravity. Find in terms of R an expression for the value of a for which the period has its minimum value. Find, again in terms of R, the expression for the minimum value of T. Check the period is indeed a minimum for the value of a you found. (You may not know how to differentiate $\sqrt{\dfrac{a^2 + \frac{1}{2}R^2}{ga}}$ but note that T will be a minimum when the function inside the square root has its minimum value.)

a) By considering the function inside the square root in the expression for T, find, in terms of R, an expression for a for which the period has its minimum value.

b) Find, in terms of R and g, an expression for the minimum value of T.

c) Find, at the value of a deduced in part a), an expression in terms of R for the second derivative of the function inside the square root in the expression for T; convince yourself that the value of the second derivative indicates that the period is a minimum at this point.

3. A battery of voltage V has an internal resistance r. It is connected across an external resistance R. The power P dissipated in the resistance R is given by $P = \left(\dfrac{V}{R+r}\right)^2 R$.

a) Assuming that V and r are constants find, in terms of r, an expression for R for which the power dissipated in the resistance R is a maximum. (You may not know how to differentiate $\left(\dfrac{V}{R+r}\right)^2 R$ but note that P will be a maximum when its reciprocal is a minimum.)

b) Using your result from part a) find, in terms of V and r, the maximum value for the power P dissipated in the resistance R.

c) Find, at the value of R deduced in part a), an expression in terms of V and r for the second derivative of the reciprocal of P; convince yourself that the value of the second derivative indicates that P is a maximum at this point.

4. The power P dissipated in a resistance R when there is a voltage V applied across it is given by $P = \dfrac{1}{R}V^2$. The resistance $R = 2.00\,\Omega$ and is known very precisely, whereas the voltage $V = 10.4 \pm 0.4$ V. Find the value of P and estimate its associated error ΔP given that there is a random measurement error ΔV of 0.4 V in V. You are asked to take two approaches to calculating the error in P which give very similar answers.

a) Calculate the value of P.

b) One estimate of the error ΔP in P resulting from the error in V is given by $\Delta P = \dfrac{dP}{dV}\Delta V$ where $\dfrac{dP}{dV}$ is evaluated at $V = 10.4$ V and $\Delta V = 0.4$ V. Find ΔP (to 2 sf) using this method.

c) Another estimate of ΔP can be obtained in the following way. Evaluate P_+, the power using $V + \Delta V = 10.4 + 0.4$ V and P_-, the value using $V - \Delta V = 10.4 - 0.4$ V. Now find your estimate for ΔP (to 2 sf) which is given by $(P_+ - P_-)/2$. (This is the average of the difference between P_+ and P_0 and between P_0 and P_-, where P_0 is the value of P when $V = 10.4$ V.)

5. The frequency f of oscillation of a mass m on a spring of spring constant k is given by $f = \dfrac{1}{2\pi}\sqrt{\dfrac{k}{m}}$. Find expressions for the following.

a) The rate of change of f with k in terms of f and k, assuming that the mass m is constant.

b) The rate of change of f with m in terms of f and m, assuming that the spring constant k is constant.

6. A particle of mass M collides head on with a stationary particle of mass m. No energy is lost in the collision. The ratio $R(m)$ of the kinetic energy of the mass m after the collision to the initial kinetic energy of mass M is given by

$$R(m) = \frac{4Mm}{(M+m)^2}.$$

a) Find, in terms of M, an expression for the value of m for which the fraction of the kinetic energy transferred to the mass m is a maximum. (You may not know how to differentiate $4Mm/(M+m)^2$ but note that R will be a maximum when its reciprocal is a minimum.)

b) Using your result from part a) find the maximum value for the fraction of the kinetic energy transferred to mass m.

c) Find, at the value of m deduced in part a), an expression in terms of M for the second derivative of the reciprocal of R; convince yourself that the value of the second derivative indicates that R is a maximum at this point.

7. The line $y = 2x + 3$ is a tangent to the curve $y = ax^2$ (where a is a constant) at a certain value of x. Find a) the value of x and b) the value of a.

8. A quadratic function has the form $y = a + bx + cx^2$ where a, b and c are constants. It has a stationary point at $(2,2)$ and, at $x = 1$, the tangent to the curve has a gradient of -2. Find the values of a, b and c. (In practice, with the information given, you will need to find b and c before finding a.)

9. The isosceles triangle shown in Fig. 3.2 has a base of length $2b$ and perpendicular height h. The length p of the perimeter of the triangle is fixed. In this question you are guided through a way of finding the maximum possible area of this triangle given that p is fixed.

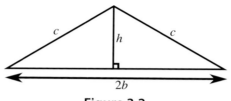

Figure 3.2

a) Write down the equation for the area A and the perimeter p of the triangle in terms of b and h. Hence, deduce an equation for A in terms of p and b.

b) Find the value of b and the corresponding value of h which maximise A.

    i. Using the equation for A you found in part a), find an expression in terms of p for the value of b which will maximise the area A of the triangle. Since p is fixed you may treat it as a constant. (You may not know how to differentiate the expression for A, but note that since A is positive it will be a maximum when A^2 is a maximum.)

    ii. Find, at the value of b deduced above, an expression in terms of p for the second derivative of A^2 with respect to b; convince yourself that the value of the second derivative indicates that A^2, and hence of A, is a maximum.

    iii. Find, in terms of p, the expression for h corresponding to this value of b.

c) Using your results from part b), find an expression for the maximum area in terms of p.

10. Consider the functions $f(x) = \dfrac{1}{x}$ and $g(x) = \dfrac{3}{2} + ax^2$. For $x > 0$ the functions have one unique point of intersection where they touch (i.e. they have the same gradient at that point).

a) Deduce the x-coordinate of the point of intersection.

b) Find the value of a.

3.5   Integration – powers, definite/indefinite integrals

In addition to the concepts from previous sections, see:
• Calculus - Introduction to Integration - Level 3:
  `isaacphysics.org/concepts/cm_integration`

1. Find the following indefinite integrals.

   a) $\int (3x - 1)(x + 1)dx.$

   b) $\int \left(\sqrt{p} - \frac{1}{p} \right)^2 dp.$

2. Find the following indefinite integrals.

   a) $\int \frac{q^2 + 3}{q^{5/2}} dq.$

   b) $\int 2z(z^2 - 1)(z^2 + 1)dz.$

3. The function $f(x)$ is such that $\int_1^3 f(x)dx = 6$ and $\int_1^5 f(x)dx = 20$. Find the value of each of the following.

   a) $\int_1^3 4f(x)dx.$

   b) $\int_1^2 f(x)dx + \int_2^5 f(x)dx.$

   c) $\int_1^3 (\frac{1}{2}f(x) - x)dx.$

   d) $\int_1^5 (f(x) - 5)dx.$

   e) $\int_3^1 f(x)dx.$

   f) $\int_3^5 f(x)dx.$

   g) $\int_1^3 (3x^2 - 4f(x))dx.$

4. Find the following integrals.

   a) $\int_1^\infty \frac{3}{2} \frac{1}{x\sqrt{x}} dx.$

   b) $\int_{-8}^0 \frac{1}{\sqrt[3]{x}} dx.$

   c) $\int_{-1}^1 \left(1 + x + \frac{x^2}{2} + \frac{x^3}{6} \right) dx.$

5. For the integrals given below, find the values of a and m respectively.

   a) $\int_{-1}^a x^{-4/5}dx = 10.$

   b) $\int_0^4 \frac{(m + 1)}{2} x^m dx = 4.$

6. A function $z(y)$ is such that $\frac{dz}{dy} = ay^3 + 4y - 1$ where a is a constant with a particular value. If $z(-1) = 0$ and $z(2) = 18$ find $z(1)$.

7. A function $v(u)$ is such that $\dfrac{dv}{du} = \dfrac{1}{3}u^{\frac{1}{3}}\left(1 - \dfrac{1}{u}\right)$ and $v(8) = -1$. Find the equation of the function.

8. A graph of the functions $y = (x-2)(x+1)$ and $y = x+1$ is shown in Fig. 3.3a. A is the dotted region between P and Q enclosed by the curve $y = (x-2)(x+1)$ and the x-axis; B is the dotted region between Q and R below the curve $y = (x-2)(x+1)$ and above the x-axis.

a) Find the area of the region A. Give your answer in the form of an improper fraction.

b) Find the area of the region B.

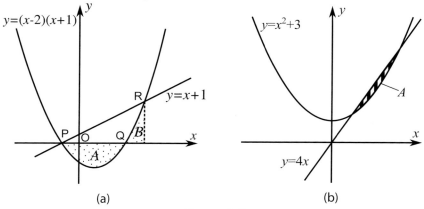

(a)   (b)

Figure 3.3

9. A graph of the functions $y = x^2 + 3$ and $y = 4x$ is shown in Fig. 3.3b. Find the area of the striped region labelled A, the region between the line $y = 4x$ and the curve $y = x^2 + 3$. Give your answer in the form of an improper fraction.

10. A graph of the functions $y = \dfrac{1}{2\sqrt{x}}$ and $y = 2x\sqrt{x}$ for $x \geq 0$ is shown in Fig. 3.4.

a) Deduce the x coordinate of the point Q.

b) Find the x coordinate of the point R.

c) Hence find the area of the dotted region OPQR, giving your answer in an exact form.

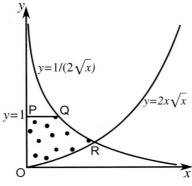

$y=1/(2\sqrt{x})$

$y=2x\sqrt{x}$

$y=1$

P   Q

R

O

x

Figure 3.4

3.6   Graph Sketching – exponentials, logs

In addition to the concepts from previous sections, see
- Graph Interpreting - Level 3:
  isaacphysics.org/concepts/cm_graph_interpreting
- Graph Sketching - Level 3:
  isaacphysics.org/concepts/cm_graph_sketching

Toolkit #3. It is helpful to think about the following when interpreting a function $f(x)$.
1. What is the main shape of the function? (linear, quadratic, cubic, reciprocal, trigonometric etc.)
2. Where are the zeros i.e. the values of x for which $f(x) = 0$?
3. What happens to the function when $x = 0$?
4. What happens to the function as x gets very large?
5. Does the function diverge anywhere within its range.
6. What are the symmetry properties of the function?
7. Does the function have minima or maxima? If so, where?
8. Does the function have points of inflection?

1. Deduce the properties of the graphs of two sets of exponential and logarithmic functions presented in Figs. 3.5 and 3.6a

   a) The graphs shown in Fig. 3.5 are exponential functions of the form $y = X^x$ and $y = X^{x+a} + b$, where X, a and b are integers. Deduce the values of X, a and b. What is the value of x at which they cross?

   b) The graphs shown in Fig. 3.6a are logarithmic functions of the form $y = \log_W x$ and $y = \log_Z(ax)$, where W, Z and a are integers. Deduce the values of W, Z and a.

2. A graph of the function $g(s)$ is shown in 3.6b. Answer the **toolkit** questions about the function $g(s)$.

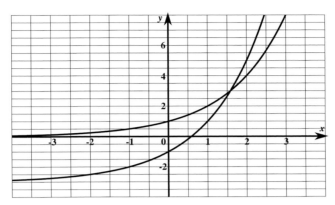

Figure 3.5

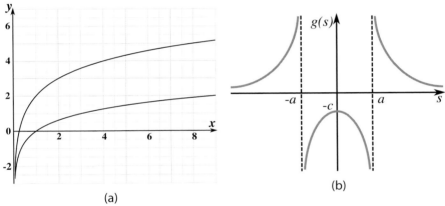

(a)

(b)

Figure 3.6

3. Consider the function $f(x) = x^3 - 3x^2 + 2$.

a) Answer the **toolkit** questions 2 to 6 about the function $f(x)$ (you are not expected to find the values of x at which the function crosses the x-axis). Hence sketch the function $f(x)$. Pay close attention to the shape of the graph and label the positions of the stationary points.

b) Use the graph you sketched in part a) to deduce the range of values of k for which the equation $x^3 - 3x^2 + 2 = k$ has three real roots.

4. Consider the functions $p(r) = 3^r - 2$ and $q(r) = \log_3(r+1)$. Answer the questions below about $p(r)$ and $q(r)$. Hence, using the same axes, sketch graphs of $p(r)$ and $q(r)$. Pay close attention to the differences in the shapes of the two functions, label the key points on both axes and indicate the positions of the asymptotes. (Do not try to work out where the two functions intersect.)

a) Answer the **toolkit** questions 2 to 5 about the function $p(r)$.

b) Answer the **toolkit** questions 2 to 5 about the function $q(r)$.

5. A graph of the potential energy $U(r)$ of a molecule a distance r from another molecule is shown in Fig. 3.7a. The force $F(r)$ between the molecules is given by the negative of the gradient of this graph i.e. $F(r) = -dU/dr$. By inspecting the graph in Fig. 3.7a answer the following questions about the behaviour of dU/dr. You will find it helpful to know that $d^2U/dr^2 = 0$ when $r = b$. Hence sketch and label a graph of $F(r)$. Pay close attention to the shape of the graph of $F(r)$ and label where the graph crosses the r-axis and the r coordinates of any stationary point .

a) What is the value of dU/dr at $r = a$?

b) How does dU/dr behave in the region $0 < r < a$?

c) How does dU/dr behave in the region $r > a$?

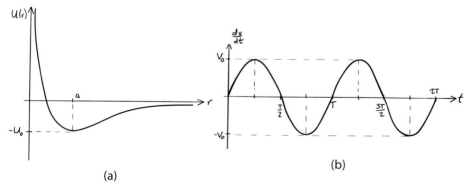

(a)

(b)

Figure 3.7

6. The displacement, velocity and acceleration of an oscillating object as a function of time t are given by $x(t)$, $v(t)$ ($= dx/dt$) and $a(t)$ ($= dv/dt$) respectively. A graph of its velocity $v(t)$ is shown in Fig.3.7b over the time range $0 \leq t \leq 2T$ where T is the period of the oscillation.

a) Deduce which of the following statements about $x(t)$ in the range $0 < t < T$ is true.

- Increasing in the range $0 < t < T/2$, decreasing in the range $T/2 < t < T$, maximum at $t = T/2$, minimum at $t = T$
- Decreasing in the range $0 < t < T/2$, increasing in the range $T/2 < t < T$, minimum at $t = T/2$, maximum at $t = T$
- Increasing in the range $0 < t < T/4$ and $3T/4 < t < T$, decreasing in the range $T/4 < t < 3T/4$, maximum at $t = T/4$, minimum at $t = 3T/4$
- Decreasing in the range $0 < t < T/4$ and $3T/4 < t < T$, increasing in the range $T/4 < t < 3T/4$, minimum at $t = T/4$, maximum at $t = 3T/4$

b) Sketch, using the graph of dx/dt in Fig.3.7b and your results from part a), the graph of $x(t)$ as a function of t over the range $0 \leq t \leq 2T$; assume that $x = 0$ when dx/dt has a maximum or minimum value, and that the maximum and minimum values of x are $\pm v_0 T/2\pi$. Pay close attention to the shape of the graph.

c) Deduce which of the following statements about the acceleration $a(t)$ in the range $0 < t < T$ is true.

- Positive in the range $0 < t < T/2$, negative in the range $T/2 < t < T$, zero at $t = T/2$ and at $t = T$
- Negative in the range $0 < t < T/2$, positive in the range $T/2 < t < T$, zero at $t = T/2$ and at $t = T$
- Positive in the range $0 < t < T/4$ and $3T/4 < t < T$, negative in the range $T/4 < t < 3T/4$, zero at $t = T/4$ and at $t = 3T/4$
- Negative in the range $0 < t < T/4$ and $3T/4 < t < T$, positive in the range $T/4 < t < 3T/4$, zero at $t = T/4$ and at $t = 3T/4$

d) Sketch, using the graph of v in Fig.3.7b and your results from part c), the graph of the acceleration of the object $a(t)$ as a function of t over the range $0 \leq t \leq 2T$. The maximum and minimum values of a are $\pm 2\pi v_0/T$. Pay close attention to the shape of the graph. How does this graph compare with the graph of $x(t)$ in part b)?

7. Answer the following questions about the function $f(x) = 2x^4 - 4x^2 + 2$. Hence sketch a graph of $f(x)$, paying close attention to the shape of the graph and labelling the positions of the stationary points.

   a) Answer the **toolkit** questions 2 to 8 about the function $f(x)$.

   b) Find the coordinates of the maximum.

   c) Find the coordinates of the minima.

8. Consider the functions $f(x) = \cos x$ and $g(x) = \dfrac{x}{4}$. Answer the questions below about $f(x)$ and $g(x)$. Hence, using the same axes, sketch graphs of the functions $f(x)$ and $g(x)$ in the range $-2\pi < x < 2\pi$.

   a) Deduce the values of $f(2\pi)$ and $f(-\pi)$.

   b) Deduce the values of $g(2\pi)$ and $g(-\pi)$ (giving your answers to 2 sf).

   c) From your graphs deduce how many solutions there are to the equation $\cos x = \dfrac{x}{4}$.

9. Consider the function $f(x) = 4\left(\dfrac{1}{x} - \dfrac{1}{x^2}\right)$. Answer the questions below about $f(x)$. Hence sketch the graph of the function $f(x)$. Pay close attention to the shape of the function, label the key points and indicate the positions of any asymptotes.

   a) Answer the **toolkit** questions 2 to 7 about the function $f(x)$.

   b) Find the coordinates of the stationary point.

10. Consider the functions $a(s) = 3\sin\left[(1-s)\dfrac{\pi}{4}\right]$ and $b(s) = \tan\left[(1-s)\dfrac{\pi}{4}\right]$. Answer the questions below about $a(s)$ and $b(s)$. Hence, using the same axes, sketch graphs of $a(s)$ and $b(s)$ in the range $-5 < s < 5$. Label the key points on both axes and indicate the positions of the asymptotes.

   a) Deduce the values of $a(-1), a(1)$ and $a(3)$.

   b) Deduce the values of $b(0), b(1)$ and $b(2)$ and find the negative value of s closest to zero at which there is a vertical asymptote.

   c) From your graph deduce how many solutions there are to the equation $3\sin\left[(1-s)\dfrac{\pi}{4}\right] = \tan\left[(1-s)\dfrac{\pi}{4}\right]$ in the range $-5 < s < 5$.

Level 4

> In addition to the concepts from previous sections, see:
> - Trigonometric Sums - Level 4:
>   isaacphysics.org/concepts/cm_trig2

1. Without using a calculator, find exact expressions for the following.

   a) $\sin 15°$.   b) $\cos 165°$.

2. Without using a calculator, find an exact expression for $\sin\left(\frac{\pi}{8}\right)$.

3. Use the addition of angles formulae to show that $\cos(90° - \theta) = \sin\theta$ and $\sin(90° - \theta) = \cos\theta$. Find the corresponding expressions for:

   a) $\cos(90° + \theta)$,   b) $\sin(90° + \theta)$.

4. If $\tan\alpha = 2$ and $\tan\beta = 0.5$, find an exact expression for $\theta = \alpha + \beta$ without using a calculator, given that $0 < \theta < \pi$.

5. Show that you can express $A\cos(\omega t + \phi)$ in the form $B\cos\omega t + C\sin\omega t$, where B and C are expressions to be found.

   a) Give an expression for B in terms of A and ϕ.

   b) Also give an expression for C in terms of A and ϕ.

6. Find the non-zero root of the equation $\sin 2\theta = (3/4)\sin\theta$ in the range from 0 to $\pi/2$, giving your answer in radians and to 3 sf.

7. Prove that $\tan 2\theta = \dfrac{2\tan\theta}{1 - \tan^2\theta}$. Now, prove that the identity $\tan 4\theta = \dfrac{k}{1 - 6\tan^2\theta + \tan^4\theta}$ is true and give an expression for k in its simplest form in terms of $\tan\theta$.

8. If $t = \tan(\theta/2)$, express the following in term of t.

   a) $\tan\theta$.   b) $\sin\theta$.   c) $\cos\theta$.

51

9. Two waves

$$\psi_1 = A \cos \left(2\pi ft - \left(\frac{2\pi}{\lambda} \right) x + \phi \right)$$

and

$$\psi_2 = A \cos \left(2\pi ft - \left(\frac{2\pi}{\lambda} \right) x - \phi \right)$$

interfere, such that the resultant wave is given by $\psi = \psi_1 + \psi_2$. Express ψ as the product of two terms.

10. Show that $X \cos \alpha t + Y \sin \alpha t$ can be expressed as $E \sin(\alpha t + \theta)$, where E and θ are expressions to be found.

a) Give an expression for E in terms of X and Y.

b) Also give an expression for θ in terms of X and Y.

4.2   Functions – e, ln, composite, modulus

In addition to the concepts from previous sections, see:
- Functions - Composite Functions - Level 4:
  isaacphysics.org/concepts/cm_functions_composite
- Trigonometric Relationships - Level 4:
  isaacphysics.org/concepts/cm_trig1

1. a) Let $f(x) = x^2 + 1$ and $g(x) = 2x + 1$. Find $g(f(x)) + f(g(x))$.

   b) Let $p(r) = 2r + 1$ and $q(r) = 3\cos r$. Find $q(p(r)) - p(q(r))$.

2. a) Let $g(a) = 2a^3$ and $h(a) = e^a$. Find $h(g(a)) + g(h(a))$.

   b) Let $y(s) = 4s^2$ and $z(s) = \ln s$. Find $z(y(s)) - 3y(z(s))$.

3. Find the following without using your calculator.

   a) $\sec\left(\dfrac{\pi}{6}\right)$.   b) $\sec(120°)$.   c) $\operatorname{cosec}\left(\dfrac{\pi}{2}\right)$.

   d) $\operatorname{cosec}\left(\dfrac{5\pi}{6}\right)$.   e) $\cot\left(\dfrac{2\pi}{3}\right)$.   f) $\cot(270°)$.

4. A Gaussian function $f(x)$ is defined to be of the form $f(x) = Ae^{-x^2/2\sigma^2}$. The response of a particular system is given by $g(z) = Be^{-z^2/2\sigma^2}e^{-hz}$. Show that this can be expressed as a Gaussian of the form $g(z) = Ce^{-(z-a)^2/2\sigma^2}$ and find expressions for the new constants a (in terms of h and σ) and C (in terms of B, h and σ).

5. γ-rays from a radioactive source pass through an arrangement consisting of n sheets of lead; the intensity of the transmitted γ-ray beam $I(n)$ as a function of n is given by $I(n) = Ae^{-\alpha n}$, where A and α are constants. Two measurements are made with q and r sheets of lead in place and the intensities are measured to be $I(q) = Q$ and $I(r) = R$.

   a) Show that A can be written in the form $(Q^r R^{-q})^\beta$ and find β.

   b) Show that α can be written in the form βS, where β is the constant derived in part a), and find an expression for S.

6. The number of radioactive particles in a specimen is given by $n = n_0 e^{-\lambda t}$.

   a) At what time has the number of particles dropped by a factor of 2?

   b) At what time has the number of particles dropped by a factor of e?

7. A particle falls through a viscous medium. Its speed $v(t)$ at time t is given by $v(t) = v_T(1 - e^{-\alpha t})$, where α and v_T are constants.

   a) What is its speed as $t \to \infty$ (the terminal velocity)?

   b) Find, in terms of α, the time taken to reach half the terminal velocity.

8. Using the fact that $\cot \phi = -3$ and that $\sin \phi$ is positive, use trigonometric identities to find exact expressions for the following.

   a) $\operatorname{cosec}\phi$.   b) $\sec \phi$.

9. a) Sketch the graph of $|\cos(2\theta)|$ in the range $-\pi \le \theta \le \pi$. How many maximums does the function have in that range?

   b) Sketch the graph of $\sin|2\theta|$ in the range $-\pi \le \theta \le \pi$. Is the function symmetric or antisymmetric about a vertical line?

   c) Solve the equation $|\cos(2\theta)| = \sin|2\theta|$ graphically in the range $0 \le \theta \le \pi$. How many solutions does the equation have in that range?.

10. a) Sketch the graph of $y = \left|\dfrac{1}{x}\right|$. Does the function diverge anywhere? Where?

    b) Sketch the graph of $y = \left|\dfrac{1}{x^2 - 4}\right|$. Does the function diverge anywhere? Where?

    c) Solve the equation $|x| = \left|\dfrac{1}{x}\right|$ graphically and give the solution as a single expression.

4.3   Series – arithmetic, geometric, binomial

In addition to the concepts from previous sections, see:
- Series - Level 4: `isaacphysics.org/concepts/cm_series`

1. Rewrite each of the following expressions in the form $a(1+b)^n$, where $|b| <$ 1. Hence, using the appropriate binomial expansion, find the value of each of them correct to 4 dp.

   a) $\sqrt{36.1}$,
   b) $\sqrt[3]{1.09}$,
   c) $\dfrac{1}{\sqrt{1.04}}$,
   d) $\sqrt[3]{125.4}$.

2. Express the following functions as partial fractions and, using the binomial expansion, find the first three terms and the general terms in their expansions in ascending powers of x. For what values of x are the expansions valid?

   a) $\dfrac{3}{(1-x)(1+2x)}$
   b) $\dfrac{(x-1)}{(x^2+2x+1)}$
   c) $\dfrac{5}{(1-x-6x^2)}$
   d) $\dfrac{(x+2)}{x^2-1}$

3. Find the sum of the arithmetic series $10.0 + 10.1 + 10.2 + \cdots + 12.0$

4. In an arithmetic progression, the fifth term is 32 and the tenth term is 57.

   a) Find the first term, a.
   b) Find the common difference, d.
   c) Hence, find the sum of the first 70 terms.

5. The first three terms of an arithmetic sequence are 6, x^2 and x.

   a) Find the possible values of x.
   b) What is the common difference, d, for the larger value of x?

6. Consider the geometric sequence $1024, -256, 64, -16, ...$

   a) Find the formula for the nth term.
   b) Find the sum of the first n terms.
   c) Find the value, to 3 sf, to which the geometric series converges.

7. The series $1 + \dfrac{1}{k} + \dfrac{1}{k^2} + \ldots$ has n terms. Find the sum.

8. The number $0.23232323\ldots$ can be written as a geometric sum with a common ratio of 0.01. Use this fact to express it as an exact fraction.

9. A bouncing ball loses the same fraction of its energy every time it bounces; i.e. when dropped from an initial height h, after its first bounce it rises to a height αh, after the second bounce ($\alpha^2 h$ and so on, α is a number less than 1). Find an expression for the total distance covered.

10. The second and fourth terms of an infinite geometric series are $\frac{1}{2}$ and $\frac{1}{72}$ respectively. Deduce the common ratio (given that it is positive) and the first term. Hence find the sum of the series, giving your answer in the form of an improper fraction.

4.4   Differentiation – e, ln, trig, chain rule, product rule

In addition to the concepts from previous sections, see:
- Calculus - Differentiation - Level 4:
  isaacphysics.org/concepts/cm_differentiation

1. a) Differentiate $a \sin \theta$ with respect to θ (a is a constant).

   b) Differentiate $A \cos(2\phi) + B$ with respect to ϕ (A and B are constants).

2. a) Find $\dfrac{ds}{d\theta}$ if $s = r \sin(\alpha\theta)$ and r and α are constants.

   b) Find $\dfrac{dq}{d\theta}$ if $q = l \cos(\alpha - 2\beta\theta)$ and l, α and β are constants.

3. The displacement x of an oscillating particle at time t is $x = A \cos(\omega t + \phi)$ where A, ω and ϕ are constants; find expressions for the velocity (the rate of change of displacement) and acceleration (the rate of change of velocity) of the particle.

4. a) Differentiate $\beta e^{-\alpha t}$ with respect to t where α and β are constants.

   b) Differentiate $Ce^{\beta m} + D$ with respect to m where β, C and D are constants.

5. a) Differentiate $3e^{4x+2}$ with respect to x.

   b) Find the rate of change of x with respect to t where $x = X(e^{\gamma t} + e^{-\gamma t})$ and X and γ are constants.

6. a) Find $\dfrac{dw}{ds}$ if $w = (4s + 3)^3$.

   b) Find $\dfrac{dz}{dw}$ when $z = (b - aw)^4$ and a and b are constants.

   c) Find $\dfrac{d^2z}{dw^2}$ when $z = (b - aw)^4$ and a and b are constants.

7. a) Find $\dfrac{du}{dv}$ and $\dfrac{d^2u}{dv^2}$ when $u = 2(3 - 2v)^{3/2}$.

   b) Find $\dfrac{dp}{da}$ if $p = \dfrac{1}{\sqrt{3a + 8}}$.

8. a) Find the (x, y) coordinates and nature of the stationary point of the function $y = (2 - 3x)^4 + 4$.

b) Consider the function $q = 4(2p - 1)^3 - 3(2p - 1)^4$. Find the stationary points of the function. How many are there? Find the values of p and q at the stationary points.

9. a) Find $\dfrac{du}{dv}$ if $u = \ln(2v + 3)$.

b) Find the coordinates and nature of the stationary point of the function $p = 2\ln(2q) - 3q$.

10. a) Find the equation of the tangent to the curve $y = e^{2x} - e^{-2x}$ at the point $x = \frac{1}{2}$.

b) Find the coordinates and nature of the stationary point of the function $u = 2e^{3v} - 3v$.

4.5   Integration – e, ln, trig

In addition to the concepts from previous sections, see:
- Calculus - Integrating Common Functions - Level 4:
  isaacphysics.org/concepts/cm_integration2

1. Find   a) $\int_0^1 2e^y dy$,   b) $\int_{-\alpha}^{\alpha} e^p dp$.

2. a) Integrate e^{-s} between $s = 0$ and $s = \infty$.   b) Find $\int_1^2 3e^{-z} dz$.

3. a) Integrate $a + be^x$, where a and b are constants, between $x = 0$ and $x = 3$.
   b) Find $\int_{-s}^{s} (e^z - e^{-z}) dz$.   c) Find $\int_{-s}^{s} (e^z + e^{-z}) dz$.

4. a) Find the indefinite integral of $4x^3$.   b) Find $\int_0^{x_0} \alpha x \, dx$ (α is a constant).

5. a) Find the indefinite integral of ax^{-8} (a is a constant).   b) Find $\int_1^2 \frac{4}{x^2} dx$.

6. a) Find $\int (ax + b) dx$, where a and b are constants.
   b) Find $\int_1^2 (x - 2)(x + 3) dx$, giving your answer as an improper fraction.

7. Find   a) $\int_{-\pi/2}^{\pi/2} 2 \cos \phi \, d\phi$,   b) $\int_{-\pi/2}^{\pi/2} \sin \alpha \, d\alpha$.

8. a) Find $\int_0^{\pi} (1 - \cos \theta) d\theta$.
   b) Find the integral of $\frac{1}{2\alpha}(2 + \sin \theta)$ between $\theta = 0$ and $\theta = \alpha$.

9. a) Find the integral of $a \cos \theta + b \sin \theta$ between $\theta = 0$ and $\frac{\pi}{3}$, where a and b are constants.
   b) Integrate $2 \cos \beta + 3 \sin \beta$ between $\beta = -\frac{\pi}{4}$ and $\beta = \frac{\pi}{4}$.

For more practice go to Section 5.4.

10. a) Find $\displaystyle\int_a^\infty \left(\frac{A}{r^7} - \frac{B}{r^{13}} \right) dr$. (The force between, for example, two atoms of an inert gas, a distance r apart is given by $\left(\frac{A}{r^7} - \frac{B}{r^{13}} \right)$, where A and B are (negative) constants; the first term is the attractive force between them (the van der Waals interaction, due to their fluctuating induced dipoles) and the second is the repulsive force due to the overlap of their electron shells. The integral describes the potential energy of such a system i.e. the work done bringing one atom from infinity to within a distance a of the other atom.)

b) Find $\displaystyle\int_{x_1}^{x_2} \left(\frac{C}{x^2} + D \right) dx$. (The function $\left(\frac{C}{x^2} + D \right)$ could describe the component of an electric field in the x-direction due to a combination of the field due to a point charge at the origin and a uniform field in the x-direction. The integral is then the potential difference between two points x_1 and x_2 on the x-axis.)

4.6   Graph Sketching – summing functions; e and ln; modulus

In addition to the concepts from previous sections, see
- Graph Interpreting - Level 4:
  `isaacphysics.org/concepts/cm_graph_interpreting`
- Graph Sketching - Level 4:
  `isaacphysics.org/concepts/cm_graph_sketching`

Toolkit #4. It is helpful to think about the following when interpreting a function $f(x)$.
1. What is the main shape of the function? (linear, quadratic, cubic, reciprocal, trigonometric etc.)
2. Where are the zeros i.e. the values of x for which $f(x) = 0$?
3. What happens to the function when $x = 0$?
4. What happens to the function as x gets very large?
5. Does the function diverge anywhere within its range.
6. What are the symmetry properties of the function?
7. Does the function have minima or maxima? If so, where?
8. Does the function have points of inflection?
9. Are there characteristic features of the function's gradient?

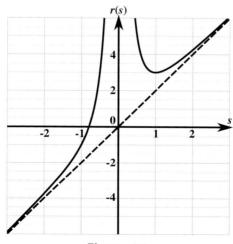

Figure 4.1

1. The graph of a function $r(s) = bs^m + cs^{-n}$, where b and c are constants and m and n are integers (where $0 \leq m < 4$ and $0 \leq n < 4$), is shown in Fig. 4.1. The dashed line shows the function $q(s) = 2s$. Deduce the values of b, c, m and n and find where $r(s) = 0$.

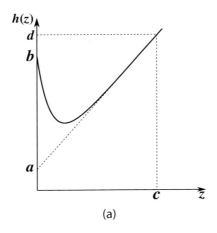

(a)

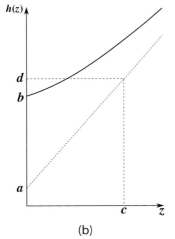

(b)

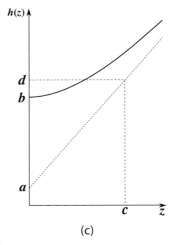

(c)

Figure 4.2

2. The graph of a function $h(z)$, which has a range $z \geq 0$, is shown in Fig. 4.2a for four positive, non-zero, constants p, q, r and s which describe the form of the function.

a) By using the information obtained from examining Fig. 4.2a, choose which one of the following functions correctly gives the form of $h(z)$:
- $h(z) = -p + qz + re^{-sz}$
- $h(z) = -p + qz + \frac{r}{z^{1+s}}$
- $h(z) = p + qz^2 + \frac{r}{z^{1+s}}$
- $h(z) = -p + qz + re^{sz}$
- $h(z) = p + qz^2 + re^{-sz}$
- $h(z) = -p + qz^2 + \frac{r}{z^{1+s}}$
- $h(z) = p + qz + re^{sz}$
- $h(z) = -p + qz^2 + re^{-sz}$
- $h(z) = p + qz + \frac{r}{z^{1+s}}$
- $h(z) = p + qz + re^{-sz}$

b) From the graph deduce an expression for p in terms of a, b, c and d (as appropriate).

c) From the graph deduce an expression for q in terms of a, b, c and d (as appropriate).

d) From the graph find an expression for r in terms of a, b, c and d (as appropriate).

e) Differentiate the expression for $h(z)$ with respect to z. Examine the behaviour of the graph at $z = 0$ and hence deduce a relationship between s and a, b, c and d.

f) Deduce a relationship between s and a, b, c and d in Fig. 4.2a.

g) Graphs with the same functional form as $h(z)$ examined in parts a) to d), but with different values of s, are shown in Figs. 4.2b and 4.2c. Examine the behaviour of each graph at $z = 0$ and hence deduce a relationship between s and a, b, c and d in each case.

3. The graph of a function $x(t) = B[\sin(2\omega t) + A\sin(\omega t)]$, where ω, A and B are constants, is shown in Fig. 4.3a. Use the hints below to deduce the values of ω, A and B.

a) You will see from examining the graph that it repeats itself after a certain value of t; what is this value of t? Given what you know about the period of sine functions deduce the value of the constant ω (give your answer in radians).

b) Find the value of the constant A. (You will find it helpful to know that the slope of the graph at $t = 0$, $t = 1$ and $t = 2$ is zero.)

c) By considering the coordinates of the stationary point labelled C find the value of B.

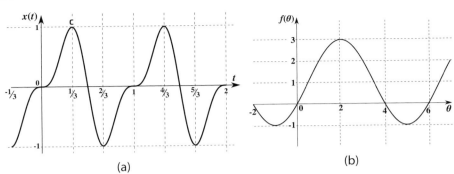

(a)

(b)

Figure 4.3

4. The graph of a function $f(\theta) = a + b^2 \sin^2(\alpha\theta + \phi)$, where a, b, α and ϕ are constants, is shown in Fig. 4.3b. Deduce the values of a, b, α and ϕ assuming that it lies in the range $0 \le \phi \le \pi$.

5. Two point objects, each of mass m, lie on the x-axis at $x = a$ and $x = -a$ respectively $(a > 0)$. The resulting gravitational potential $V(x)$ on the x-axis is given by $V(x) = -\dfrac{A}{|x - a|} - \dfrac{A}{|x + a|}$ where A is a positive constant.

Answer the questions below about the behaviour of $-\dfrac{A}{|x - a|}$. Hence, on the same axes sketch graphs of the two functions $-\dfrac{A}{|x - a|}$ and $-\dfrac{A}{|x + a|}$; then, again on the same axes, sketch $V(x)$ by considering the sum of the two functions at each point. From your graph and the symmetry of $V(x)$ deduce the coordinates of the stationary point of $V(x)$.

a) Answer the **toolkit** questions 2 to 8 about the function $f(x) = -\dfrac{A}{|x - a|}$.

b) From your graph deduce the coordinates of the stationary point of $V(x)$.

6. A mass on a spring is displaced by a distance A from its equilibrium position and released from rest at time $t = 0$. Its motion is heavily damped so that it does not oscillate and its subsequent displacement x as a function of time t is given by $x(t) = \frac{4}{3}Ae^{-\gamma t} - \frac{1}{3}Ae^{-4\gamma t}$ where γ is a positive constant. Using the hints given below sketch how the displacement and velocity of the mass vary as a function of time t, and find the time at which the velocity has its largest (negative) value.

a) To examine how the displacement $x(t)$ behaves as a function of t, sketch the two functions $\frac{4}{3}Ae^{-\gamma t}$ and $-\frac{1}{3}Ae^{-4\gamma t}$ on the same graph in the range $0 \le t \le 4/\gamma$; then, again on the same graph, sketch $x(t)$ by considering the sum of the two functions at each point. Note that since the mass is released from rest, its velocity is zero (i.e. $dx/dt = 0$) at $t = 0$. You will also find it helpful to know that $e^{-1} = 0.37$ and $e^{-4} = 0.018$.

b) By differentiating the expression for the displacement of the mass $x(t)$ with respect to t find an expression for the velocity $v(t)$ of the mass at time t.

c) Find the time at which v has its largest (negative) value.

d) Find the largest (negative) value of $v(t)$.

e) Using the results from part b) sketch the graph of $v(t)$ as a function of t in the range $0 \le t \le 4/\gamma$. (You will have found that $v(t)$ is the sum of two functions; you may find it helpful to sketch these two functions on the same graph and then consider the sum of the two functions at each point.)

7. A rocket is launched vertically from the surface of the Earth. The mass of the rocket is M and initially it is carrying a mass m of fuel. Burnt fuel is ejected at a constant speed u relative to the rocket.

a) Ignoring the effects of gravity the final speed v of the rocket (when all the fuel has been consumed) as a function of m/M is $v = u \ln \left(1 + \frac{m}{M}\right)$. Sketch a graph of v as a function of m/M (i.e. the horizontal axis is the ratio m/M).

b) In one particular case, the mass of the fuel is ten times the mass of the rocket (i.e. $m = 10M$). Burnt fuel is ejected at a constant speed u relative to the rocket, and at a constant rate $\mu = 10M/T$, where T is the time taken for the fuel to be used. Ignoring the effects of gravity the speed $v(t)$ of the rocket at time t during the burn is $v(t) = -u \ln \left(1 - \frac{\mu t}{11M}\right) = -u \ln \left(1 - \frac{10t}{11T}\right)$. Show that $v(T/2)$ can be written in the form αu where α is a number; find α giving your answer to 2 sf.

c) $v(T)$ can be written in the form βu where β is a number; find β giving your answer to 2 sf.

d) Sketch a graph of $v(t)$ in the range $0 \le t \le T$. You may find it helpful to note that the expression $\left(1 - \frac{10t}{11T}\right)$ is less than 1 and to recall what the graph of $\ln(x)$ is like for $x < 1$

8. In this question you are asked to sketch the function $\cos 2\theta$ and hence, using the double angle formula for cosine, to sketch the graphs of $\sin^2 \theta$ and
$\cos^2 \theta$.

a) Consider $f(\theta) = \cos 2\theta$. Find the value of $f(\pi/4)$ and $f(\pi/2)$.

b) With the help of the results in part a) sketch the graph of $\cos 2\theta$ in the
range $0 < \theta < 2\pi$. Label both axes suitably.

c) Use the identity $1 - \cos 2\theta = 2\sin^2 \theta$, and your sketch from part b), to
sketch the graph of $\sin^2 \theta$ in the range $0 < \theta < 2\pi$. Label both axes suitably.

d) On the sketch you produced in part c), sketch the graph of $\cos^2 \theta$ in the
range $0 < \theta < 2\pi$. You may find it helpful to use your sketch from part b)
and the identity $1 + \cos 2\theta = 2\cos^2 \theta$.

9. Consider the function $r(s) = \frac{1}{2}(s - \alpha)^2 + \frac{1}{2}(s + \alpha)^2$ where α is a constant.
Answer the questions below about the behaviour of $\frac{1}{2}(s - \alpha)^2$. Hence on
the same axes sketch graphs of the two functions $\frac{1}{2}(s - \alpha)^2$ and $\frac{1}{2}(s + \alpha)^2$
in the range $-3\alpha \leq s \leq 3\alpha$; then again on the same axes, sketch $r(s)$ by
considering the sum of the two functions at each point.

a) Answer the **toolkit** questions 2 to 8 about the function $g(s) = \frac{1}{2}(s - \alpha)^2$.

b) Find the s component of the stationary point of $g(s) = \frac{1}{2}(s - \alpha)^2$.

c) Find the value of $g(s)$ at the stationary point.

10. A particle falls from rest through a viscous medium. Its speed $v(t)$ at time t is
given by $v(t) = v_T(1 - e^{-\alpha t})$ where v_T and α are constants. Its corresponding position x below its starting point at time t (obtained by integrating this
expression and assuming $x(0) = 0$) is given by $x(t) = v_T \left(t - \frac{1}{\alpha}\left(1 - e^{-\alpha t}\right) \right)$.

a) Find the value of $v(0)$.

b) Find the value of $v(t)$ as $t \to \infty$.

c) With the help of the results in parts a) and b) sketch a graph of the speed
$v(t)$ of the particle, as a function of t, for $t \geq 0$.

d) The position of the particle below its starting point is given by $x(t) =$
$v_T \left(t - \frac{1}{\alpha}\left(1 - e^{-\alpha t}\right) \right)$. Sketch a graph of $x(t)$ as a function of time t for $t \geq$
0. You will find it helpful to sketch the functions $v_T \left(t - \dfrac{1}{\alpha} \right)$ and $v_T \dfrac{e^{-\alpha t}}{\alpha}$
on the same graph and then to add them point by point to sketch $x(t)$.

Level 5

In addition to the concepts from previous sections, see:
- Vectors - Resolving Vectors - Level 5:
  isaacphysics.org/concepts/cm_vectors2

1. Find the angle between the vectors $a = \hat{i} + 2\hat{j} + 4\hat{k}$ and $b = 2\hat{i} + 3\hat{j} + \hat{k}$. Give your answer to 3 sf.

2. The angle between the vectors p and q is $60°$, $|q| = 5$ and $p \cdot q = 10$. What is the length of p ?

3. Find the scalar product $a \cdot b$, where $a = \hat{i} + 2\hat{j} + 4\hat{k}$ and $b = 2\hat{i} - 3\hat{j} + \hat{k}$. Hence, deduce the angle between a and b. Give your answer to 3 sf.

4. Find $r \cdot s$, where $r = \hat{i} + 3\hat{j} + 2\hat{k}$ and $s = -\hat{i} - \hat{j} - 2\hat{k}$.

5. a) Write down the scalar product of the vectors u and v where $u = u_x\hat{i} + u_y\hat{j} + u_z\hat{k}$ and $v = v_x\hat{i} + v_y\hat{j} + v_z\hat{k}$.

   b) From your result in part a) deduce an expression for the square of the length of the vector u which is given by $u^2 = |u|^2 = u \cdot u$.

6. In the triangle of Fig. 5.1a $c = b - a$.

   a) Find, in terms of a and b, an expression for $c \cdot c$.

   b) From your result in part a) and using the symbols shown in the diagram prove the cosine formula for a triangle.

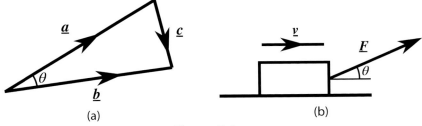

(a)   (b)

Figure 5.1

67

7. A large block is being pulled at a constant velocity $v = 5.00$ m s^{-1} across a horizontal floor by an applied force $F = 122$ N directed at an angle $\theta = 37.0°$ above the horizontal as shown in the diagram of Fig. 5.1b. What is the rate at which the force does work on the block? (The rate of doing work is given by $F \cdot v$.)

8. A force $F = 2\hat{i} + 3\hat{j} + \hat{k}$ N moves a particle from $r_1 = \hat{i} + 2\hat{j}$ m to $r_2 = 2\hat{i} + 4\hat{j} - \hat{k}$ m.

   a) Find the work done by the force (the work done on the particle is given by $F \cdot r$ where r is its displacement).

   b) Find the component of F in the direction of the displacement (to 3 sf).

9. A force $F = F_x\hat{i} + F_y\hat{j} + F_z\hat{k}$ does work on a particle moving it through a displacement $\Delta r = \Delta x\hat{i} + \Delta y\hat{j} + \Delta z\hat{k}$.

   a) Find a general expression for the work done by the force (given by $F \cdot \Delta r$).

   b) Consider the case where $F_x = F_0 \cos\theta$, $F_y = F_0 \sin\theta$ and $F_z = 0$.

      i. If $\Delta x = a$, $\Delta y = 0$ and $\Delta z = 0$, what is the work done?

      ii. If $\Delta x = a \sin\theta$, $\Delta y = -a \cos\theta$ and $\Delta z = 0$, what is the work done?

10. A particle is moving in the (x, y)-plane in a circle of radius r centred on the origin such that $r^2 = x^2 + y^2$; its displacement from the origin at any time is given by $r = x\hat{i} + y\hat{j}$. The force on the particle is given by $F = \dfrac{F_0}{r}(y\hat{i} - x\hat{j})$.

   a) Find the magnitude of the force F.

   b) Find the angle between the force F and the displacement r of the particle from the origin. Give your answer to 3 sf.

   c) How much work (given by $F \cdot r$) is done by the force as the particle moves round?

5.2   Functions – rational, polynomials

In addition to the concepts from previous sections, see:
- Functions - Polynomials and Rational Functions - Level 5:
  isaacphysics.org/concepts/cm_functions_polynomials

1. Simplify $\dfrac{p+2}{p-3} + \dfrac{p(1-p)}{p^2-2p-3}$.

2. Simplify $\dfrac{b+1}{2b-1} - \dfrac{3b-1}{4b^2-1}$.

3. Simplify $\dfrac{y-2}{2(y^2-y)} \times \dfrac{3y^2+y-4}{16-4y^2}$.

4. Simplify $\dfrac{x-1}{x+2} + \dfrac{3-x}{x+1}$.

5. Simplify $\dfrac{p^2-q^2}{p^2-pq-6q^2} \times \dfrac{p+2q}{2p^2+pq-q^2}$.

6. Simplify $\dfrac{r^2+7rs+10s^2}{4s^2-4rs-3r^2} \div \dfrac{5s^2+16rs+3r^2}{2s^2+3rs-9r^2}$.

7. The function $\dfrac{2x-1}{(3x-2)(x-1)}$ can be written as $\dfrac{A}{3x-2} + \dfrac{B}{x-1}$. Find A and B.

8. Split the function $\dfrac{x}{(x+2)(x+3)}$ into partial fractions and hence find

$$\int_0^1 \dfrac{x}{(x+2)(x+3)}dx.$$

9. Write the function $\dfrac{2z^2-z-3}{(z+2)(z^2-2z-1)}$ in the form $\dfrac{A}{z+2} + \dfrac{B+Cz}{z^2-2z-1}$,

finding A, B and C. Hence, find $\displaystyle\int_1^2 \dfrac{2z^2-z-3}{(z+2)(z^2-2z-1)}dz.$

10. The function $\dfrac{w+2}{(w-1)(w+1)(2w+1)}$ can be written as $\dfrac{A}{(w-1)} + \dfrac{B}{(w+1)} +$

$\dfrac{C}{(2w+1)}$. Using the substitution method find the constants A, B and C.

For more practice go to Section 6.2.

5.3   Differentiation – implicit, chain rule, product rule

In addition to the concepts from previous sections, see:
- Calculus - Differentiation- Level 5:
  isaacphysics.org/concepts/cm_differentiation

1. Find the following.

   a) $\dfrac{dp}{dt}$ where $p = (4t^2 + 3)^{-3}$.

   b) $\dfrac{dp}{dq}$ where $p = \dfrac{1}{[(q+1)^2 + (q-1)^2]}$.

2. Find the following.

   a) $\dfrac{dE}{dt}$ if $E = B\sin^2(\omega t)$.   b) $\dfrac{dy}{dx}$ if $y = e^{-x^2/(2\sigma^2)}$.

3. a) Use the product rule to find the derivative w.r.t t of $(t+1)(3-t^2)$.

   b) Find $\dfrac{ds}{dt}$ if $s = \dfrac{t}{(1+t^3)}$.

4. a) Find $\dfrac{dy}{dt}$ where $y = at^2 e^{\beta t}$.

   b) Find the derivative w.r.t. θ of $\tan\theta$ by writing it as $\dfrac{\sin\theta}{\cos\theta}$.

5. a) Use the product rule to find the derivative w.r.t. x of $\dfrac{f(x)}{h(x)}$ and thus prove the quotient rule.

   b) Use the result for the derivatives of inverse functions, i.e. $\dfrac{dy}{dx} = 1/\left(\dfrac{dx}{dy}\right)$, to find the derivative of $y = \ln x$. Hence, using this result, find $\dfrac{df}{dr}$ if $f = r\ln r$.

6. a) Find the derivative w.r.t. x of $x^2 \ln \left(\dfrac{k}{x} \right)$.

   b) Find the value of a which maximises V when $V = Aa \ln \left(\dfrac{b}{a} \right)$ (with A and b positive constants) and deduce the corresponding value of V.

7. a) Find $\dfrac{dy}{dx}$ if $x^2 + y^2 = r^2$. Give your answer as a function of x and y.

   b) Find, as a function of x and y, the gradient of the tangent to the curve $x^2 - xy + y^2 = 7$.

   c) Using the equation for the gradient of the tangent to the curve $x^2 - xy + y^2 = 7$ from part b) evaluate the slope at the point $(-1, 2)$.

8. a) One modification to the perfect gas equation of state $(pV = RT)$ which takes account of the finite sizes of the molecules and intermolecular attractions is Van der Waals' equation, given by $(p + \dfrac{a}{V^2})(V - b) = RT$, where a and b are constants. For a gas obeying Van der Waals' equation, find an expression for $\dfrac{dp}{dV}$ assuming T is a constant. Give your answer as a function of a, b, R, p and V only.

   b) Another modification to the perfect gas equation of state which takes account of the finite sizes of the molecules and intermolecular attractions is Dieterichi's equation, given by $p(V - b) = RTe^{-a/RTV}$, where a and b are constants. For a gas obeying Dieterich's equation, $p(V - b) = RTe^{-a/RTV}$. Find an expression for $\dfrac{dV}{dT}$ assuming p is a constant.

9. a) Find the coordinates and nature of the stationary points of the function $y = xe^{-x^2/2}$. How many are there?

   b) The displacement x of a damped, simple harmonic oscillator as a function of time t is given by $x = Ae^{-\alpha t} \cos(\omega t + \phi)$. Find a general expression for the times at which the displacement is a maximum or minimum. (Your answer will involve an inverse trig function and an arbitrary integer n.)

10. a) A particle of mass m_1 undergoes a head-on elastic collision with a stationary particle of mass m. The particle of mass m then collides elastically and head-on with a third stationary particle of mass m_2. The fraction of the original kinetic energy of m_1 passed on to m_2 is given by:

$$E = \frac{4m_1 m_2 m^2}{[(m + m_1)(m + m_2)]^2}.$$

   i. Find an expression for the value of m which maximises E.

   ii. Find the value of E which corresponds to the value of m found in part a.i), giving your answer in terms of r where $r = \sqrt{\dfrac{m_1}{m_2}}$.

b) The amplitude A of a damped harmonic oscillator when being forced to oscillate at a frequency ω is given by $A = \dfrac{B}{((\omega_0^2 - \omega^2)^2 + 4\gamma^2\omega^2)^{1/2}}$ where ω_0, γ and B are constants. (ω_0 is the frequency of free undamped oscillations of the system, γ is related to the amount of damping and B is related to the amplitude of the applied force).

   i. Find an expression for the value of ω for which A takes its maximum value (ω is positive).

   ii. Determine the value of A corresponding to the value of ω found in part b.i). (This is called amplitude resonance).

5.4   Integration – by parts, substitution, trig identities

In addition to the concepts from previous sections, see:
- Calculus - Integrating Common Functions - Level 5:
  isaacphysics.org/concepts/cm_integration2
- Calculus - Integration by Substitution and by Parts - Level 5:
  isaacphysics.org/concepts/cm_integration3

Find the following integrals.

1. a) $\int \sin(c\theta)d\theta.$

   b) $\int e^{av}dv.$

2. a) $\int (bv + c)^2 dv.$

   b) $\int_0^b a(y - b)^3 dy.$

3. a) $\int_0^2 \frac{3}{(z+1)^2}dz.$

   b) $\int \frac{e^{-ax}}{(1 + e^{-ax})^4}dx.$

4. a) $\int_0^a \frac{1}{b}\frac{1}{(x+a)}dx.$

   b) $\int_0^1 \frac{x}{1+x^2}dx.$

5. a) $\int_a^{2a} \frac{x^3}{a^5 + ax^4}dx.$

   b) $\int_0^{\pi/4} \tan \beta \, d\beta$ (by writing $\tan \beta = \sin \beta / \cos \beta$).

6. a) $\int \frac{1}{\sqrt{1 - bz^2}}dz.$

   b) $\int A\cos(\omega t)\cos(3\omega t)dt.$

7. a) $\int_0^{2\pi} \frac{1}{2\pi} \sin^2 \theta d\theta.$

   b) $\int_0^L \cos^2\left(\frac{\pi x}{L}\right) dx.$

8. a) $\int_0^\alpha (s+1)e^{-s/\alpha}ds.$

   b) $\int_0^{\pi/2} x \sin x dx.$

9. a) $\int z^2 \cos z dz.$

   b) $\int_1^2 3t^2 \ln t dt.$

10. a) $\int_1^3 \frac{2}{x}dx.$

    b) $\int_a^b \frac{A}{s}ds$ (where $a, b > 0$).

5.5　Differential Equations – first order

In addition to the concepts from previous sections, see:
• Differential Equations - Level 5:
  isaacphysics.org/concepts/cm_differential_equations

1. Find the solution to the differential equation $\dfrac{dp}{dq} = pq$ given that $p = 2$ when $q = 0$.

2. Find the solution to the differential equation $\dfrac{da}{db} = \dfrac{2b}{a^3}$ given that $a = 1$ when $b = \dfrac{1}{2}$.

3. The velocity, $\dfrac{dx}{dt}$, of a particle undergoing simple harmonic motion is given by the equation $\dfrac{dx}{dt} = \omega A \cos(\omega t + \phi)$ where x is the displacement at time t, and ω, A and ϕ are constants. Solve this differential equation to obtain the equation for x as a function of t given that $x = 0$ at $t = 0$.

4. The velocity, $\dfrac{dq}{dt}$, of a particle undergoing simple harmonic motion is given by the equation $\dfrac{dq}{dt} = \omega(a^2 - q^2)^{1/2}$ where q is the displacement at time t, and ω and a are constants. Solve this differential equation to find the equation for q as a function of t, given that $q = a$ at $t = 0$.

5. The velocity and displacement of a particle at time t are given by v and x respectively. The acceleration of a particle $\dfrac{dv}{dt}$ can, via the chain rule, be written as $v\dfrac{dv}{dx}$. The force on the particle is given by $-kx$ and its mass is m. Use Newton's Second Law to write down the differential equation relating v and x; solve the equation, given that $v = 0$ when $x = x_0$.

6. A rocket has a mass m and velocity v at time t; it accelerates by ejecting some of its mass (in the form of gas produced by combustion of fuel inside the rocket) at a speed v_0 relative to the rocket. Using conservation of momentum, the equation relating m and v is $\dfrac{dm}{dv} = -\dfrac{m}{v_0}$. By solving the differential equation find an expression for the velocity of the rocket when it has a total mass M_2, given that it starts from rest with a mass M_1.

7. Find the solution to the differential equation $\dfrac{du}{dy} = -Aye^{-\alpha y^2}$ given that $u \to 0$ as $y \to \infty$.

8. The acceleration of a particle is given by $-bv$ where v is the velocity at time t and b is a constant. The particle has an initial velocity v_0.

   a) Given that acceleration is the rate of change of velocity with time write down the differential equation describing the system and solve this to find an expression for v in terms of v_0, b and t.

   b) Now use the fact that velocity is equal to the rate of change of displacement x to write down the differential equation relating x and t. Solve this equation to find how x varies with t given that $x = 0$ when $t = 0$.

9. A capacitor (of capacitance C) and a resistor (of resistance R) are in series with a battery; the switch in the circuit is open and the capacitor is uncharged. When the switch is closed the rate at which the charge q on the capacitor increases with time t is given by $\dfrac{dq}{dt} = \dfrac{Q_0}{RC}e^{-t/RC}$. Solve this differential equation to find the equation for the charge q as a function of time t, given that $q \to Q_0$ as $t \to \infty$.

10. The rate at which a body cools is given by $-C\dfrac{dT}{dt}$ where T is its temperature at time t and C is its heat capacity (the minus sign indicates its temperature is decreasing as time increases; i.e. it is cooling). If T_0 is the temperature of the surroundings the rate of cooling is given by $k(T - T_0)$ where k is a constant. Write down the relevant differential equation and solve it to find the equation for the temperature T as a function of time t given that the temperature is T_1 at $t = 0$.

5.6   Graph Sketching – products of functions; algebraic functions

In addition to the concepts from previous sections, see
- Graph Interpreting - Level 5:
  isaacphysics.org/concepts/cm_graph_interpreting
- Graph Sketching - Level 5:
  isaacphysics.org/concepts/cm_graph_sketching

Toolkit #5. It is helpful to think about the following when interpreting a function $f(x)$.
1. What is the main shape of the function? (linear, quadratic, cubic, reciprocal, trigonometric etc.)
2. Where are the zeros i.e. the values of x for which $f(x) = 0$?
3. What happens to the function when $x = 0$?
4. What happens to the function as x gets very large?
5. Does the function diverge anywhere within its range.
6. What are the symmetry properties of the function?
7. Does the function have minima or maxima? If so, where?
8. Does the function have points of inflection?
9. Are there characteristic features of the function's gradient?

1. The graph of a function $w(z)$, which has a range $z \geq 0$ and which is itself the product of two functions $u(z)$ and $v(z)$, is shown in Fig. 5.2a. The form of the function $w(z)$ is described by three positive, non-zero, constants A, α and β.

a) Use the information obtained from examining the graph to deduce the form of $w(z)$. In the expressions given below the three constants A, α and β are positive and non-zero.

- $w(z) = A \sin(\alpha z)e^{\beta z}$
- $w(z) = A \sin(\alpha z)e^{-\beta z}$
- $w(z) = A \cos(\alpha z)z^{-\beta}$

- $w(z) = A \cos(\alpha z)e^{\beta z}$
- $w(z) = A \cos(\alpha z)e^{-\beta z}$
- $w(z) = A \sin(\alpha z)z^{-\beta}$

b) By considering the form of $w(z)$ you deduced in part a), find the value of A (give your answer to 2 sf).

c) By considering the form of $w(z)$ you deduced in part a), find the value of α (give your answer to 2 sf).

d) By reading off the value of $w(z)$ for an appropriate value of z, deduce the value of β (give your answer to 2 sf).

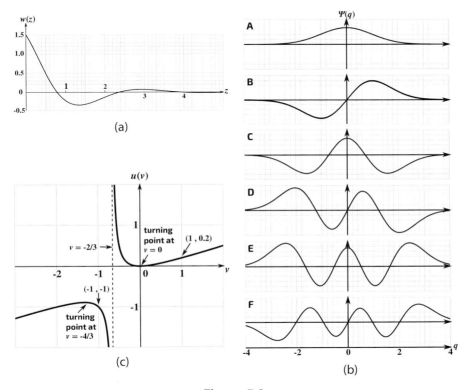

Figure 5.2

2. The six graphs shown in Fig. 5.2b are the products of a set of what are called Hermite polynomials and an exponential function of the form $e^{-q^n/2}$ where n is an integer in the range $0 \leq n \leq 3$. (To do this question you do not need to understand the physical significance of these functions. However, if you are interested they represent what are called the wavefunctions $\psi(q)$ of a quantum harmonic oscillator (see chapter 3 of "A Cavendish Quantum Mechanics Primer" on isaacphysics.org/qmp) for different values of its energy; q is proportional to the displacement of the oscillator from its equi-

librium position, and the square of the wavefunction (i.e. $(\psi(q))^2$) gives the probability of finding the oscillator at that value of q.)

a) Deduce the value of n.

b) Which of the graphs is described by an equation of the form $\psi(q) = A(q - aq^3)e^{-q^n/2}$ where a is a constant?

c) The energy of a quantum harmonic oscillator is given by $(m + \frac{1}{2})E_0$ where m is the number of zeros in its wave function $\psi(q)$ and E_0 is a constant.

   i. What is the value of m for the function $\psi(q) = A(q - aq^3)e^{-q^n/2}$?

   ii. Which of the wave functions represents the state with the lowest energy?

   iii. Which of the wave functions represents the state with the highest energy?

d) Deduce from the graph describing the function $\psi(q) = A(q - aq^3)e^{-q^n/2}$ the values of q at which the probability of finding the oscillator is zero.

e)   i. How many points of inflection are there in the function $\psi(q) = A(q - aq^3)e^{-q^n/2}$?

   ii. In this case a point of inflection at which $\psi(q)$ is not zero is the value of q which marks the point beyond which the system could not exist in classical physics (it would have negative kinetic energy in this region!). How many of these points are there in the function $\psi(q) = A(q - aq^3)e^{-q^n/2}$ and at what values of q do they occur?

3. The graph of a function $u(v) = \dfrac{v^n}{pv + q}$, where p, q and n are integers, is shown in Fig 5.2c. Deduce the values of p, q and n.

a) From the graph deduce the ratio $\dfrac{q}{p}$.

b) From the information given on the graph in Fig. 5.2c deduce the value of p. (You will find it helpful to note that there is one value of v for which the value of $u(v)$ does not depend on n.)

c) Deduce the value of q.

d) Find a formula for the turning points of the function $u(v)$. Hence, using the information you are given about the graph in Fig. 5.2c, deduce the value of n.

4. The graph of a function $p(q) = \dfrac{c}{q^2 + aq + b}$, where a, b and c are integers, is shown in Fig. 5.3. Deduce the values of a, b and c.

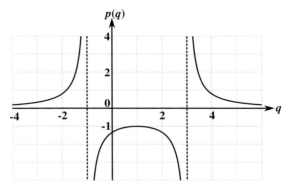

Figure 5.3

5. Consider the function $f(t) = e^{-t^2}\cos(6\pi t)$. Answer the following questions about the functions e^{-t^2} and $\cos(6\pi t)$. Hence, on the same axes sketch graphs of these two functions in the range $-1 \le t \le 1$. By considering the product of these two functions at each point, sketch, on a separate graph, the function $f(t) = e^{-t^2}\cos(6\pi t)$, also in the range $-1 \le t \le 1$.

a) What is the value of e^{-t^2} at $t = 1$? Give your answer to 2 sf.

b) Find the value of $\dfrac{d(e^{-t^2})}{dt}$ at $t = 0$.

c) The function e^{-t^2} has two points of inflection; find the values of t at which these occur.

d) How many times, in the range $-1 \le t \le 1$, is $\cos(6\pi\theta) = 1$?

e) How many times, in the range $-1 \le t \le 1$, is $\cos(6\pi\theta) = -1$?

6. Consider the functions $q(r) = 4 - r^2$ and $p(r) = |8 - 2r^2|$. Answer the following questions about $q(r)$ and $p(r)$. Hence, sketch the function $q(r) = 4 - r^2$ in the range $-3 \le r \le 3$ and, on the same graph, in the same range, sketch the function $p(r) = |8 - 2r^2|$.

a) Consider the function $q(r) = 4 - r^2$. Find the value of $q(0)$ and $q(-3)$ and the values of r at which q = 0.

b) Consider the function $p(r) = |8 - 2r^2|$. Find the value of $p(0)$ and $p(-3)$.

7. Consider the function $h(x) = Ax^2e^{-\alpha x^2}$ where α and A are positive constants. Answer the following questions about the function $h(x)$. Hence sketch the function $h(x)$. (You may find it helpful to sketch the functions Ax^2 and $e^{-\alpha x^2}$ separately on the same graph, and then sketch the function $h(x)$ by considering the product of these two functions at each point.)

a) Answer the **toolkit** questions 2 to 8 about the function $h(x)$.

b) How many stationary points does the function $h(x)$ have?

c) Find the coordinates of the stationary point of $h(x)$ with the lowest value of x. What is the nature of this stationary point?

d) Find the coordinates of the stationary point of $h(x)$ with the second lowest value of x. What is the nature of this stationary point?

e) From your examination of the behaviour of the function $h(x)$ in parts a), b), c) and d) deduce how many points of inflection $h(x)$ has.

8. Consider the function $z(y) = \dfrac{-9}{(y^2 + 2y - 8)}$. Answer the questions below about this function and hence sketch a graph of $z(y)$.

a) Answer the **toolkit** questions 2 to 8 about the function $z(y)$.

b) Find the coordinates of the stationary point of $z(y)$.

9. Consider the function $q(p) = \dfrac{1}{p^2 - p}$. Answer the questions below about the function $q(p)$ and hence sketch a graph of $q(p)$. Use your graph to deduce the range of values of c for which there are no real solutions to the equation $\dfrac{1}{x^2 - x} = c$.

a) Answer the **toolkit** questions 2 to 8 about the function $q(p)$ and hence sketch a graph of $q(p)$.

b) By examining the graph in part a) deduce the range of values of c for which there are no real solutions to the equation $\dfrac{1}{x^2 - x} = c$.

10. An oscillator consisting of a mass m on a spring is subject to viscous damping. The mass is displaced from its equilibrium position and released from rest at time $t = 0$. At a later time t its speed is $v(t) = -v_0 e^{-t/\tau} \sin(\omega t)$ where v_0, τ and ω are positive constants. Its kinetic energy $T(t)$ at time t is given by $T(t) = Ae^{-2t/\tau} \sin^2(\omega t)$ where $A = mv_0^2/2$. Assume that $m = 0.40$ kg, $v_0 = 0.30$ m s^{-1}, $\omega = \pi$ s^{-1} and $\tau = 4.0$ s. Answer the following questions about the functions $e^{-t/\tau}$ and $- \sin(\omega t)$. Hence sketch, on the same axes, graphs of the two functions $e^{-t/\tau}$ and $- \sin(\omega t)$ in the range $t = 0$ s to $t = 4.0$ s. By considering the product of these two functions at each point, sketch, on a separate graph, the function $v(t)$, also in the range $t = 0$ s to $t = 4.0$ s. On the same axes, and by examining the relationship between $v(t)$ and $T(t)$, sketch a graph of $T(t)$.

a) What is the value of $e^{-t/\tau}$ when $t = \tau/2$ and when $t = \tau$? Give your answers to 2 sf.

b) Give a general expression, in terms of ω, π and n (where n is any integer), for the times at which $- \sin(\omega t) = 0$ and for the times at which $- \sin(\omega t) = 1$.

c) Find the value of v and T when $t = 0.50$ s.

5.7   Advanced Algebraic Manipulation – on-line

Additional practice of algebra in interesting physical contexts.

See isaacphysics.org/assignment/pre_uni_maths_lvl5_7

Level 6

6.1   Vectors – vector products

> In addition to the concepts from previous sections, see:
> • Vectors - Vector (or Cross) Products - Level 6:
>   isaacphysics.org/concepts/cm_vectors3

1. a) Deduce the directions of: $\hat{k} \times \hat{i}$, $\hat{j} \times \hat{k}$, $\hat{k} \times \hat{j}$ and $-\hat{i} \times \hat{k}$.

   b) The vector $b = c \times d$ where $c = c_x\hat{i} + c_z\hat{k}$ and $d = d_x\hat{i} + d_y\hat{j}$. The vector b can be written as $b = b_x\hat{i} + b_y\hat{j} + b_z\hat{k}$. Find b_x, b_y and b_z.

2. a) The vector $r = p \times q$ where $p = -\hat{i} + 5\hat{k}$ and $q = 2\hat{k}$. The vector r can be written as $r = r_x\hat{i} + r_y\hat{j} + r_z\hat{k}$. Find r_x, r_y and r_z.

   b) The vector $c = a \times b$ where $a = \hat{i} + \hat{j} + \hat{k}$ and $b = -\hat{i} + \hat{j} + \hat{k}$. The vector c can be written in the form $c = c_x\hat{i} + c_y\hat{j} + c_z\hat{k}$. Find c_x, c_y and c_z.

3. a) A force $F = (2\hat{i})$ N acts at a point with position vector $r = (2\hat{i} + 3\hat{j})$ m; find the magnitude and direction of the torque, $\tau (= r \times F)$, that the force F produces about the origin.

   b) A force $F = F_x\hat{i} + F_y\hat{j}$ acts at a point with position vector $r = x\hat{i} + y\hat{j} + z\hat{k}$; the torque $\tau (= r \times F)$ that the force F produces about the origin is given by $\tau = \tau_x\hat{i} + \tau_y\hat{j} + \tau_z\hat{k}$. Find expressions for τ_x, τ_y and τ_z.

4. A force F acts in the xy-plane at a point P which lies on the x-axis a distance a from the origin (as in Fig. 6.1a); F makes an angle of θ to the x-axis as shown in the diagram. Find the magnitude and direction of the torque ($= r \times F$) produced by F about the origin (the force F has a magnitude of F).

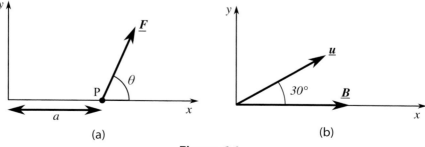

(a)   (b)

Figure 6.1

5. a) Find the magnitude and direction of the force acting on a particle with a charge $q = 1.6 \times 10^{-19}$ C moving with velocity $v = (1.0 \times 10^6 \hat{\imath})$ m s^{-1} in a magnetic field $B = (2.0 \times 10^{-4}\hat{k})$ T.

b) A particle with a charge $q = 3.2 \times 10^{-19}$ C is moving in the xy-plane with a velocity u of 2.0×10^6 m s^{-1} at an angle of $30°$ to the x-axis. A magnetic field B of magnitude 3.0×10^{-4} T lies along the x-axis as shown in Fig. 6.1b. Find the magnitude and direction of the force acting on the particle.

6. A particle is rotating about the x-axis with angular velocity $w = (10\hat{\imath})$ rad s^{-1}. Its velocity v ($= w \times r$) when it is at the point $r = (\hat{\jmath} + \hat{k})$ m is given by $v = v_x\hat{\imath} + v_y\hat{\jmath} + v_z\hat{k}$. Find v_x, v_y and v_z.

7. A particle is rotating with angular velocity w about an axis through the origin. Its position vector at time t is r, and its velocity v at that time is given by $v = w \times r$. If $w = \Omega\hat{k}$ and $r = r\cos\theta\hat{\imath} + r\sin\theta\hat{\jmath}$, find the following.

a) v.   b) The scalar product $v \cdot r$.

c) Deduce the angle between v and r.

8. a) A particle of charge $+q$ is moving with velocity v in a magnetic field $B = B_0\hat{k}$; the force on the particle is given by $F = F_0\hat{\jmath}$. Find the magnitude and direction of the particle's velocity v.

b) A particle with charge $-q$ is moving with velocity $v = v_0\hat{\jmath}$ in a magnetic field $B = B_0\hat{\imath}$. Find the magnitude and direction of the force F on the particle.

9. a) The vector $c = a \times b$ where $a = 4\hat{\imath} + \hat{\jmath}$ and $b = \hat{\imath} + 4\hat{\jmath}$. Find c. Hence find the area of the parallelogram with sides a and b. The angle between the vectors a and b is ϕ, in degrees; find ϕ, giving your answer to the nearest degree.

b) The vector area A of the parallelogram with sides $p = \hat{\imath} + \hat{\jmath} + \hat{k}$ and $q = 2\hat{\imath} + 3\hat{\jmath}$ is equal to $p \times q$. The vector A can be written in the form $A = A_x\hat{\imath} + A_y\hat{\jmath} + A_z\hat{k}$; find A_x, A_y and A_z.

10. a) A long straight wire lies along the y-axis and carries a current I flowing in the positive y-direction; it lies in a region where there is a magnetic field B. Write down expressions for the magnitude and direction of the force on length Δl of the wire when (a) $B = B_0\hat{i}$, (b) $B = B_0\hat{j}$, (c) $B = B_0\hat{k}$.

b) A wire is carrying a current of $1.0\,\text{mA}$ flowing in the positive x-direction. A uniform magnetic field B of $4.0 \times 10^{-5}\,\text{T}$ is at angle of $60°$ to the wire in the xy-plane as shown in Fig.6.2. Find the magnitude and direction of the force per unit length on the wire. The direction of the magnetic field is reversed; what is the direction of the force on the wire now?

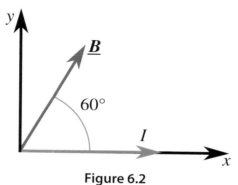

Figure 6.2

For more practice go to Section 7.1.

6.2   Functions – hyperbolic, sinc

In addition to the concepts from previous sections, see:
- Functions - Hyperbolic Functions - Level 6:
  `isaacphysics.org/concepts/cm_functions_hyperbolic`

1. Show that a) $\sinh 2x = 2\sinh x \cosh x$, b) $\cosh 2x = \cosh^2 x + \sinh^2 x$.
   c) Using the previous results find the expression for $\tanh 2x$ in terms of $\tanh x$.

2. a) Show that the derivative of $\text{sech}\, x$ is $\alpha\, \text{sech}\, x \tanh x$ and find α.
   b) Show that the derivative of $\coth x$ is $\beta\, \text{cosech}^2 x$ and find β.

3. Find a) $\int_{-\infty}^{\infty} \dfrac{1}{\cosh^2(x)}\,dx$, b) $\int_{-\infty}^{\infty} \text{sech}^4(x)\,dx$.

4. A uniform chain of mass m and length l is sliding off a smooth table; at time
   t length x hangs over the edge of the table, see Fig. 6.3a. Show that the
   equation of motion is $m\dfrac{d^2x}{dt^2} = \dfrac{mgx}{l}$ where g is the acceleration due to
   gravity. Show, by substitution that $x = A\cosh(\lambda t) + B\sinh(\lambda t)$, where A
   and B are constants, is a solution if λ satisfies a particular algebraic equation.

   a) Find the expression for λ, assuming that it is positive.

   b) At $t = 0$ a length b of chain is hanging over the edge of the table and the
   chain is at rest, find i) A and ii) B.

   c) Find, in terms of λ, l and b, the speed of the chain when $x = l$.

   d) Find the final acceleration.

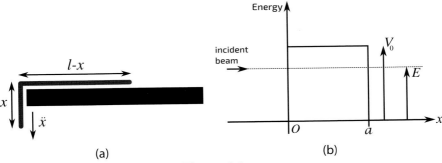

(a)   (b)

Figure 6.3

5. A uniform chain of mass m and length l is sliding off a rough table; at time t length x lies over the edge of the table (see 6.3a) with enough weight to drag the chain downwards; the acceleration due to gravity is g and the coefficient of kinetic friction μ_k. The equation of motion is of the form

$$m\frac{d^2x}{dt^2} = \frac{mx}{l}g - \mu_k\frac{m(l-x)}{l}g,$$

where the first term on the righthand side is the force downwards due to the weight of the chain hanging over the edge of the table and the second is the frictional force between the chain and the table.

Rearrange the equation of motion into the form

$$\frac{d^2x}{dt^2} - \alpha x = \beta$$

where α and β are constants which depend on the acceleration due to gravity g, the coefficient of kinetic friction μ_k and l.

a) Find the expressions for i) α and ii) β

b) The general solution of the equation above is given by $x(t) = x_c(t) + x_p(t)$, where x_c is the complementary function (general solution of the homogeneous equation) and x_p is a particular integral (the solution of the inhomogeneous equation). Substitute the values of α and β you found in part a) into the inhomogeneous equation and hence derive an expression for x_p in terms of g, μ_k and l.

c) Show by substitution into the homogeneous equation that the equation $x_c = A\cosh(\lambda t) + B\sinh(\lambda t)$, where A and B are constants, is a solution if λ satisfies a certain algebraic equation. Find λ in terms of g, μ_k and l.

d) At $t = 0$ a length b of chain is lying over the edge of the table and the chain is at rest. Use these initial conditions to find expressions for the constants A and B in Part C in terms of b, μ_k and l.

6. A beam of particles, each of mass m and with total energy E, is travelling in the positive x-direction and is incident on a potential barrier of width a; the height of the barrier V_0 is greater than E (i.e. $V_0 > E$), and it is of width a as shown in Fig. 6.3b. Classically particles will not pass through such a barrier; however in quantum mechanics particles can pass through this classically

forbidden region (where $E < V_0$). This process is called tunnelling. The fraction T of particles that pass through the region is given by:

$$T = \frac{4k^2\alpha^2}{[(k^2 - \alpha^2)^2 \sinh^2(\alpha a) + 4k^2\alpha^2 \cosh^2(\alpha a)]}$$

where $k = \sqrt{2mE/\hbar^2}$ and $\alpha = \sqrt{(2m(V_0 - E))/\hbar^2}$, and $\hbar = h/2\pi$ where h is Planck's constant.

a) Show that T can be written as $T = \dfrac{1}{A \sinh^2(\alpha a) + 1}$. Find A.

b) Using your result from part a) and recalling the definition of sinh in terms of exponentials show that, if $\alpha a \gg 1$, then $T \approx Be^{-2\alpha a}$. Find B.

7. A particle wave in one-dimension can be described by a wavefunction $\psi(x)$, which we assume specifies the state of the particle entirely. $\psi(x)$ when squared can be interpreted as giving the probability density $P(x)$ of finding the particle at position x; density in this case means the probability per unit length, so $P(x)dx$ gives the probability of finding a particle between x and $x + dx$. The total probability of finding the particle in the region from $x = -\infty$ to $x = \infty$ must be 1 so $\int_{-\infty}^{\infty} P(x)dx = \int_{-\infty}^{\infty} \psi^2(x)dx = 1$. The wavefunction $\psi(x)$ is a solution of the Schrödinger equation:

$$-\frac{h^2}{8\pi^2 m}\frac{d^2\psi}{dx^2} + V(x)\psi(x) = E\psi(x)$$

where m is the mass of the particle, $V(x)$ is the potential energy of the particle as a function of x, E is its total energy. The particle is confined to a one-dimensional potential given by $V(x) = -V_0 \operatorname{sech}^2(x/\sigma)$ where $V_0 = \hbar^2/(m\sigma^2)$. Show by substitution that, for a suitable k value, $\psi(x) = A/\cosh(kx)$ is a solution of the Schrödinger equation for this potential. [This problem is posed, with its full quantum mechanical implications, in "A Cavendish Quantum Mechanics Primer", Exercise 4.34, available in the same series as this book on isaacphysics.org/qmp.]

a) Find an expression for k.

b) Find an expression for the energy E in terms of σ, h and m.

c) Evaluate the normalisation constant A.

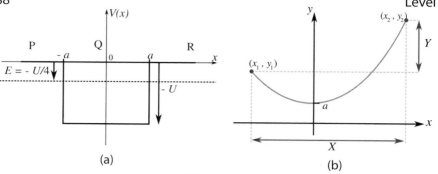

Figure 6.4

8. Consider now the situation in which the particle from question 6 is confined to a one-dimensional square potential well of width $2a$ and depth $-U$ and has a total energy $E = -U/4$ as shown in Fig. 6.4a. In the regions P and R, outside the potential well, the particle's total energy is less than its potential energy; these regions, where – according to classical physics – the kinetic energy of the particle is negative, are described as being classically forbidden. For this value of the energy the solutions of the Schrödinger equation in the regions labelled P, Q and R are given by $\psi_P(x) = -Be^{\alpha x}$ for $x < -a$, and $\psi_Q(x) = A\sin(kx)$ for $-a \leq x \leq a$ and $\psi_R(x) = Be^{-\alpha x}$ with $x > a$. In this question you are asked to find the probability of finding the particle outside the well in the classically forbidden region.

a) To find the value of k, use the Schrödinger equation above with the value of $V(x)$ in region Q and the value of E; substitute the given expression for $\psi_Q(x)$ and deduce an expression for k. Repeat this operation for region R to deduce an expression for α. Hence find the ratio k/α.

b) To match the wavefunctions at the boundary between the regions Q and R the wavefunction and its gradient must be continuous at $x = a$, that is: $\psi_Q = \psi_R$ and $\dfrac{d\psi_Q}{dx} = \dfrac{d\psi_R}{dx}$. Use these conditions to find the value of ka given that $0 \leq ka \leq \pi$. (Similar conditions apply at the boundary between the regions P and Q but do not give any added information in this case.)

c) Use one of the conditions you found at the boundary and the value of ka you deduced in part b) to find an expression in terms of αa for the ratio A/B.

d) By considering the probabilities of finding the particle in each of the areas P, Q and R calculate the probability that the particle is outside the well, in the classically forbidden regions P and R. Show that it can be written in the form $\dfrac{C}{8\pi + 12\sqrt{3}}$, and find an exact expression for the constant C.

9. Monochromatic light of wavelength λ is incident normally on a slit of width a. The resulting diffraction pattern is displayed on a screen; the angle between the line from the slit to a point on the screen and the normal to the screen is θ. The intensity $I(\theta)$ of the diffraction pattern at an angle θ is given by

$$I(\theta) = I_0 \left[\text{sinc} \left(\frac{ka \sin \theta}{2} \right) \right]^2 \quad \text{where } k = \frac{2\pi}{\lambda}, \ I_0 \text{ is a constant and the func-}$$

tion $\text{sinc}\beta = \left(\frac{\sin \beta}{\beta} \right)$.

a) Find the value of $\text{sinc}\beta$ when $\beta = 0$.

b) Using part a), find the intensity of the central primary maximum at $\theta = 0$.

c) There are zeros in intensity either side of the central primary maximum for certain angles θ_n, where n is a non-zero integer; the zeros closest to the centre are θ_1 and θ_{-1}, the next closest are θ_2 and θ_{-2} etc. Find an expression in terms of n, λ and a for the values of $\sin \theta_n$ at which the intensity is zero.

d) Between the zeros in intensity there are subsidiary maxima. By considering the function $\text{sinc}\beta$ find an equation for β from which the positions of these maxima can be calculated. (Note that the transcendental equation you obtain for β, which involves a trig function, cannot be solved analytically.)

e) As discussed in part c), either side of the central primary maximum at $\theta = 0$ there are zeros in intensity for certain angles θ_n, where n is a non-zero integer; the zeros closest to the centre are θ_1 and θ_{-1}, the next closest are θ_2 and θ_{-2} etc. Between these minima there are the subsidiary maxima whose positions could be found from the equation you derived in part d).

   i. Find the ratio of the intensity of the subsidiary maximum between the two zeros at θ_1 and θ_2 and the intensity of the primary maximum.

   ii. Find the ratio of the intensity of the subsidiary maximum between the two zeros at θ_2 and θ_3 and the intensity of the central primary maximum. (To answer this, you will require the coordinates of the first two non-zero solutions of the transcendental equation obtained in part d); these occur when $\beta = 1.43\pi$ and $\beta = 2.46\pi$.)

10. It can be shown that an inextensible chain or rope supported at both ends and hanging freely in a uniform gravitational field takes up the shape called a catenary which has the form $y = a\cosh(x/a)$; using this notation the vertex of the catenary lies on the y-axis at height a as shown in Fig. 6.4b. The parameter a is determined by the length of the chain or rope L, the horizontal separation of the supports X and the difference in the heights of the two ends Y (see Fig. 6.4b); the coordinates of the tops of the supports are (x_1, y_1) and (x_2, y_2). The equations for L and Y are

$$L = a\sinh\left(\frac{x_2}{a}\right) - a\sinh\left(\frac{x_1}{a}\right)$$

$$Y = y_2 - y_1 = a\cosh\left(\frac{x_2}{a}\right) - a\cosh\left(\frac{x_1}{a}\right)$$

Consider $L^2 - Y^2$; eliminate x_1 and x_2 to obtain an equation relating L, Y, X and a. (Given a set of values of L, Y and X this equation can only be solved numerically for a.)

6.3   Series – Maclaurin, Taylor

In addition to the concepts from previous sections, see:
- Series - Levels 5 and 6: isaacphysics.org/concepts/cm_series

1. a) Expand e^{x^2} up to x^4.   b) Expand $e^{-x^2/2\sigma^2}$ up to x^4.

2. a) i. Write down the Maclaurin expansion of $\ln(1+z)$ up to z^3.
   ii. By re-writing $\ln(2+4y)$ in the form $A + \ln(1+z)$, where A is a constant, find the Maclaurin expansion of $\ln(2+4y)$ up to y^3.
   b) Find the first 4 non-zero terms in the expansion of $\ln(1+q) - \ln(1-q)$.

3. Expanding e^x, find the first three non-zero terms in the Maclaurin series for $\operatorname{sech} x$.

4. a) Expand $Ae^{-\alpha t}$ up to the term in t^2.

   b) Find the first two non-zero terms in the expansion of $e^p - e^{-p}$.

   c) A lightly damped oscillatory system has a period T. The total energy of the system at time t is given by $E(t)$. One period later its energy $E(t+T) = E(t)e^{-\gamma T}$.

   i. Find an expression for the fractional change in energy in one cycle.
   ii. On the assumption that $\gamma T \ll 1$ find an approximate expression for the fractional change in energy in one cycle.

5. a) Find the value of $e^{-0.1}$ to 4 decimal places without using a calculator.

   b) Find the value of $e^{0.2}$ to 3 dp without using a calculator.

6. a) Find, using a Maclaurin expansion, the cosine of 0.2 rad, to 3 dp.

   b) Find, using a Maclaurin expansion, the sine of 0.08 rad, to 2 sf.

   c) A pendulum consists of a point mass m suspended on a light string of length l. When the string makes an angle of ϕ to the vertical its potential energy relative to the point where $\phi = 0$ is given by $mgl(1 - \cos\phi)$. Show that for $\phi \ll 1$ the potential energy is given approximately by $A_0\phi^2$ and find an expression for A_0.

7. a) Write down the third non-zero term in the expansion of $\sin(4\theta)$.

   b) Using the standard trigonometric formula for the cosine of the sum of two angles write $\cos((\pi/3) - \alpha)$ in terms of $\cos\alpha$ and $\sin\alpha$. Hence find the first 5 terms in the Maclaurin expansion of $\cos((\pi/3) - \alpha)$.

8. a) Find the first two non-zero terms in the Maclaurin expansion of $\cos\phi$.

   b) Using your result from part a) and the Binomial expansion find the first two non-zero terms in the series for $1/\cos\phi \equiv (\cos\phi)^{-1}$.

   c) Hence, using $\tan\phi = \sin\phi/\cos\phi = \sin\phi(\cos\phi)^{-1}$, multiply the result from part b) and the Maclaurin expansion of $\sin\phi$ to get the first two non-zero terms in the Maclaurin expansion of $\tan\phi$.

   d) Using the fact that $\sin(2\alpha) = 2\sin\alpha\cos\alpha$, multiply the Maclaurin expansions of $\cos\alpha$ and $\sin\alpha$ together to find the first three non-zero terms in the Maclaurin expansion of $\sin(2\alpha)$. Now find the Maclaurin series for $\sin(2\alpha)$ directly and verify that the first three non-zero terms in the series are the same as the previous part of the question. Find the α^7 term.[1]

9. a) An electric dipole consists of two charges $+q$ and $-q$ separated in the z direction by a very small distance a. The electric potential $V(z)$ a distance z away from the centre of the dipole in a direction along the line joining the two charges is given by:

$$V(z) = \frac{q}{4\pi\epsilon_0}\left(\frac{1}{z-(a/2)} - \frac{1}{z+(a/2)}\right).$$

Find an approximate expression for $V(z)$ (assume that $z \gg a$ and obtain the first non-zero term in the Maclaurin expansion of $V(z)$).

   b) An electric dipole consists of two charges $+q$ and $-q$ separated by a distance a. The electric potential V a distance r from the centre of the dipole in a direction making an angle θ to the line joining the two charges is given approximately by:

$$V \approx \frac{q}{4\pi\epsilon_0}\left(\frac{1}{\sqrt{r^2 - ar\cos\theta}} - \frac{1}{\sqrt{r^2 + ar\cos\theta}}\right).$$

Assuming that $r \gg a$, show that $V \approx A\cos\theta/r^2$ and find A.

10. The potential energy of two atoms a distance r apart interacting via a van der Waals force is given by $V(r) = U_0\left(\left(\frac{a}{r}\right)^{12} - 2\left(\frac{a}{r}\right)^{6}\right)$. Show that this function has a minimum at $r = a$. Use a Taylor expansion about $r = a$ to obtain an approximate expression for the potential at a point $h = r - a$ to second order in h.

---

[1] Python resources for this book, including Maclaurin expansions, can be found at
https://isaacmaths.org/book/python_exercises

6.4   Differentiation – inverse functions, chain rule, product rule

1. a) Find $\dfrac{dt}{ds}$ if $t = \tfrac{1}{3}\log_4(s^2)$, giving your answer in its simplest possible form.

   b) Differentiate $\ln(uv) = ue^v$ to show that $\dfrac{dv}{du}$ can be written as:

   $$\frac{dv}{du} = \frac{f(u,v)\,v}{u\,(1 - uve^v)}$$

   where $f(u,v)$ is a function of u and v, and find an expression for $f(u,v)$.

   c) Differentiate $q = (p+1)^{2(p+1)}$ to show that the derivative with respect to p of q can be written as $\dfrac{dq}{dp} = g(p)(p+1)^{2(p+1)}$ where $g(p)$ is a function of p, and find an expression for $g(p)$.

2. a) Find $\dfrac{db}{da}$ where $b = A\sin^{-1}a$ and A is a constant. The range of the function is $-\dfrac{\pi}{2} \le \dfrac{b}{A} \le \dfrac{\pi}{2}$.

   b) Using your result from part a) find $\dfrac{dq}{dp}$ where $q = \sin^{-1}(p/3)$.

   c) Using your result from part a) find $\dfrac{d\phi}{d\alpha}$ where $\phi = \beta\ln(\sin^{-1}\alpha)$, β is a constant and $0 < \sin^{-1}\alpha < \pi/2$.

3. a) Find $\dfrac{dw}{dv}$ where $w = \cos^{-1}v$. The range of the function is $0 \le w \le \pi$.

   b) Using your result from part a) find $\dfrac{dh}{dg}$ where $h = \dfrac{1}{2}(\cos^{-1}g)^2$. The range of the function is $0 \le h \le \dfrac{\pi^2}{2}$.

4. a) Find $\dfrac{ds}{dr}$ where $s = R\tan^{-1}r$ and R is a constant. The range of the function is $-\dfrac{\pi}{2} \le \dfrac{s}{R} \le \dfrac{\pi}{2}$.

   b) Find $\dfrac{dy}{dx}$ where $y = A\sec^{-1}\left(\dfrac{a}{x}\right)$ and A and a are constants. The range of the function is $0 \le \dfrac{y}{A} \le \pi$.

5. a) Find $\dfrac{dz}{dy}$ if $z = \sinh^{-1} y$.

b) Find $\dfrac{d\psi}{d\theta}$ if $\psi = a \operatorname{sech}^{-1}(\cos b\theta)$ where a and b are positive constants and $0 < b\theta < \pi/2$.

6. A particle wave in one-dimension can be described by a wavefunction $\psi(x)$, which we assume specifies the state of the particle entirely. For a particle in a quadratic potential well the wavefunction $\psi(x)$ is a solution of the Schrödinger equation:

$$-\frac{h^2}{8\pi^2 m}\frac{d^2\psi(x)}{dx^2} + \frac{1}{2}kx^2\psi(x) = E\psi(x)$$

where m is the mass of the particle, $\dfrac{1}{2}kx^2$ is the potential energy of the particle as a function of position x, E is the total energy of the particle and the quantity h is Planck's constant. In parts a) to e) of this question you are guided through a way in which this equation can be solved.

a) Use the chain rule to show that, for an appropriate value of α you can rewrite the Schrödinger equation in the form:

$$\frac{d^2 W(z)}{dz^2} + (\lambda - z^2)W(z) = 0$$

where $z = \alpha x$ and $\psi(x) = W(z)$. Find an expression for the constant α in terms of h, k and m.

b) From your analysis in part a) find λ in terms of E, h, k and m.

c) The solution to the equation $\dfrac{d^2 W(z)}{dz^2} + (\lambda - z^2)W(z) = 0$ can be written in the form $W(z) = y(z)e^{-z^2/2}$. Show $y(z)$ is a solution of the equation $\dfrac{d^2 y(z)}{dz^2} + f(z)\dfrac{dy(z)}{dz} + cy(z) = 0$ where $f(z)$ is a function of z and c is a constant. Find $f(z)$.

d) From your analysis in part c) find an equation, in terms of λ, for the constant c.

e) You found $f(z)$ and c in parts c) and d). Show, by substituting the expressions into the differential equation for $y(z)$ (given in part c)), that $y_1(z) = z$

and $y_2(z) = 1 + az^2$ (where $a \neq 0$) are both solutions of this equation and find the corresponding values of λ_1 and λ_2 and the value of a. (Note that there are only certain allowed solutions of the Schrödinger equation for a particle in a quadratic potential. From your expression for λ in part b), it is possible to deduce the unique value for the total energy E associated with each of the different solutions for $y(z)$ and the corresponding values of λ; the total energy E of the system can only take these values and is therefore described as being quantized.)

7. A hydrogen atom consists of a single electron with charge $-q$ in orbit around a nucleus with charge $+q$. The electrostatic potential of the electron in the field of the nucleus is spherically symmetrical and is given by $-q^2/(4\pi\epsilon_0 r)$ where r is the distance of the electron from the nucleus. A spherically symmetric quantum state of the electron is given by the wavefunction $\psi(r)$ which relates to the probability of finding the electron at a given value of r. For this spherically symmetric state the wavefunction $\psi(r)$ is a solution of the Schrödinger equation:

$$-\frac{h^2}{8\pi^2 m}\frac{1}{r^2}\frac{d}{dr}\left(r^2\frac{d\psi(r)}{dr}\right) - \frac{q^2}{4\pi\epsilon_0 r}\psi(r) = E\psi(r)$$

where m is the mass of the particle, E is the total energy of the particle and the quantity h is Planck's constant. In the ground (lowest energy) state of the hydrogen atom $\psi(r) = Ae^{-r/a_0}$, where a_0 is a constant known as the Bohr radius and A is called the normalisation constant. Find, by substituting this expression for $\psi(r)$ into the Schrödinger equation, expressions for a_0 and E in terms of h, m, q and ϵ_0.

8. A quantum harmonic oscillator has quantised energy levels which are equally separated in energy ϵ, i.e. the energy levels are $0, \epsilon, 2\epsilon,...n\epsilon,...$ To investigate its thermal properties it is convenient to use something called the partition function Z where $Z = \sum_{n=0}^{\infty} e^{-n\beta\epsilon}$ and $\beta = 1/kT$ (k is the Boltzman constant and T is the absolute temperature of the system). For a system of N oscillators it can be shown that the average energy U is given by $U = -N\frac{1}{Z}\frac{dZ}{d\beta}$; the heat capacity C is $C = \frac{dU}{dT}$. In part a) of this question you are asked to find a simple expression for Z, in parts b) and c), to deduce general expressions for U and C, and in parts d) and e), to find the values of U and C in the high temperature limit when $kT \gg \epsilon$.

a) Find the sum of the series $Z = \sum_{n=0}^{\infty} e^{-n\beta\epsilon}$.

b) Using the expression for Z deduced in part a), find an expression for U in terms of N, ϵ and β, with the term in β appearing in the denominator only.

c) Remembering that $\beta = 1/kT$, use the expression for U you derived in part b) to obtain an expression for the heat capacity C in terms of N, k, ϵ and T.

d) In the high temperature limit $kT \gg \epsilon$. Using your result from part b) find an expression for U in terms of N, k and T in the high temperature limit.

e) In the high temperature limit $kT \gg \epsilon$. Using your result from part c) find an expression for C in terms of N and k in the high temperature limit.

9. It can be shown that for blackbody radiation at temperature T the energy $U_f(f)$ per unit volume per unit frequency range at frequency f is given by

$$U_f(f) = \frac{8\pi h f^3}{c^3(e^{hf/kT} - 1)}$$ where h is the Planck constant, c is the speed of

light and k is the Boltzmann constant. The energy $U_\lambda(\lambda)$ per unit volume per unit wavelength range at wavelength λ is given by $U_\lambda(\lambda) = -U_f(f)\dfrac{df}{d\lambda}$ where $f\lambda = c$. The minus sign arises because an increase in frequency corresponds to a decrease in wavelength. In this question you are guided through the derivation of the Wien displacement law, $\lambda_m T = b$, where λ_m is the wavelength at which U_λ is a maximum at temperature T and b is a constant.

a) By writing f in terms of λ in the expression for $U_f(f)$ and using the relationship given above between $U_\lambda(\lambda)$ and $U_f(f)$ show that $U_\lambda(\lambda)$ can be written in the form $U_\lambda(\lambda) = g(\lambda)\dfrac{1}{e^{hc/\lambda kT} - 1}$, where $g(\lambda)$ is a function of λ. Find an expression for the function $g(\lambda)$, in terms of λ, h and c.

b) $U_\lambda(\lambda)$ will be a maximum when $\dfrac{dU_\lambda}{d\lambda} = 0$. Show that this occurs when $e^{-x} + w(x) = 0$, where $x = \dfrac{hc}{\lambda kT}$ and $w(x)$ is a simple function of x. Find $w(x)$.

c) There is no analytical solution to the equation $e^{-x} + w(x) = 0$ derived in part b); however it can be shown numerically that $x = 4.965$ is a solution to 4 sf. To check this evaluate $e^{-4.965} + w(4.965)$ giving your answer to 1 sf.

d) From the solution for x in part c) deduce an expression for the constant b in the Wien displacement law.

10. The number $g(N, m)$ of ways of sharing m quanta each of the same energy amongst N oscillators is given by $g(N, m) = \dfrac{(N + m - 1)!}{(N - 1)!\, m!}$. Two independent systems, system 1 and system 2, contain N_1 and N_2 oscillators respectively, where N_1 and N_2 are fixed. The systems are in thermal contact, sharing m quanta between them, such that system 1 has m_1 quanta and system 2 has m_2 quanta; the total number of quanta m ($= m_1 + m_2$) is fixed. The number of ways in which this can be done is given by:

$$g(N, m) = \frac{(N_1 + m_1 - 1)!}{(N_1 - 1)!\, m_1!} \frac{(N_2 + m_2 - 1)!}{(N_2 - 1)!\, m_2!}$$

where $N = N_1 + N_2$. In part a) of this question you are asked to find the most probable value of m_1 (when g is a maximum), and in part b), to show that g is indeed a maximum at this value of m_1. You may assume that N_1, N_2, m_1 and m_2 are all $\gg 1$ so that you can use Stirling's approximation for the factorial terms here, i.e. $\ln(n!) = n \ln n - n$.

a) Using Stirling's approximation find an expression for $\ln g$. Take the derivative with respect to m_1 of the resulting expression for $\ln g$; hence find the value of m_1 for which the function g has its largest value, giving your answer in terms of m, N_1 and N. Remember that $m = m_1 + m_2$ is a constant.

b) Use the expression for $\dfrac{d \ln g}{dm_1}$ you found in part a) to obtain $\dfrac{dg}{dm_1}$ and hence $\dfrac{d^2 g}{dm_1^2}$. Show that, at the stationary point found in part a), the second derivative can be written in the form $g_m \dfrac{X}{(N + m)m}$ where g_m is the value of g at the stationary point (you do not need to find an expression for g_m itself). Obtain an expression for X in terms of N, N_1 and N_2 and convince yourself that g_m is indeed a maximum.

   Additional question

11. Consider the function $f(x) = x^{1/x}$.

   a) Find the Taylor expansion to second order of this function about $x = e$.

   b) Hence find an expression for $f(\pi)/f(e)$ to second order in $(\pi - e)$, and from this deduce which is bigger, $e^{1/e}$ or $\pi^{1/\pi}$.

6.5   Integration – by parts, partial fractions, substitution

1. Find $\int \dfrac{\ln x}{x^2} dx$.

2. Find $\int \dfrac{\ln(\sin x)}{\cos^2 x} dx$.

3. Find $\int_{-\infty}^{\infty} Ax^2 e^{-x^2/2\sigma^2} dx$, where A is a constant, given that
   $\int_{-\infty}^{\infty} Ae^{-x^2/2\sigma^2} dx = 1$. (The expression inside the second integral,
   $Ae^{-x^2/2\sigma^2}$, is the Gaussian or Normal Distribution and describes the probability density for the variable x which has a mean of 0 and a standard devi-
   ation of σ; the integral of this expression from $-\infty$ to ∞ is 1 as the variable
   x must lie in this range. The first integral will give the mean value of x^2.)

4. Find, by integrating by parts twice, $\int_0^{\pi/3} e^{-x} \sin x dx$.

5. Find $\int \dfrac{(x+3)}{(x+2)(x+4)} dx$.

6. Find $\int_0^a \dfrac{1}{a^2 + x^2} dx$ using the substitution $x = a \tan \theta$.

7. Find $\int_1^{\infty} \dfrac{4}{(e^x - e^{-x})^2} dx$.

8. Find, using the method of partial fractions, $\int_0^b \dfrac{1}{(a+x)(b+x)} dx$.

9. Find an expression for y if $\int_0^y \dfrac{1}{\sqrt{u^2 - \omega^2 x^2}} dx = t$. (This describes the motion of a simple harmonic oscillator, where ω is the frequency of the oscillator, y is its displacement from equilibrium at time t and u is its maximum
   speed.)

10. Find $\int \dfrac{1}{4 + 5 \cos \theta} d\theta$, using the substitution $t = \tan(\theta/2)$.

6.6   Differential Equations – first order, second order

In addition to the concepts from previous sections, see:
• Differential Equations - Level 6:
  isaacphysics.org/concepts/cm_differential_equations

1. Find the solution of the equation $2y\dfrac{dy}{dx} = y^2 + x$ given that $y = 0$ when $x = 0$.

2. Find the solution of the equation $\dfrac{1}{v}\dfrac{du}{dv} = \dfrac{4u}{v^2} + 2$ given that $u = 0$ when $v = 1$.

3. Find the general solution of the equation $x\dfrac{dy}{dx} + (a + x)y = e^{-x}$.

4. The equation describing the displacement x of the bob of a damped pendulum from its equilibrium position is $\dfrac{d^2x}{dt^2} = -\omega_0^2 x - 2\gamma\dfrac{dx}{dt}$, where ω_0 is the angular frequency of undamped oscillations of the pendulum and γ is related to the damping. Assuming $\omega_0 > \gamma$ find an equation for x at time t given that $x = X$ and $\dfrac{dx}{dt} = 0$ at $t = 0$. (You will find it helpful to define a new constant ω_1 such that $\omega_1^2 = \omega_0^2 - \gamma^2$.)

5. A radioactive element (1) is produced by the decay of another radioactive element (2). The differential equation describing this process is $\dfrac{dn}{dt} = \lambda_2 N - \lambda_1 n$, where n is the number of nuclei of element 1 at time t, N is the number of nuclei of element 2 at time t and λ_1 and $\lambda_2(\lambda_1 > \lambda_2)$ are related to the decay rates of elements 1 and 2 respectively. Assuming that $N = N_2 e^{-\lambda_2 t}$ the equation can be arranged to become $\dfrac{dn}{dt} + \lambda_1 n = \lambda_2 N_2 e^{-\lambda_2 t}$. Find how the number n of nuclei of element 1 varies with time t given that $n = N_1$ at $t = 0$.

6. Find the solution of the equation $\dfrac{d^2p}{dq^2} - 4\dfrac{dp}{dq} + 3p = 3q - 1$ given that $p = 2$ and $dp/dq = -1$ when $q = 0$.

7. The equation of motion of a forced oscillator is given by the differential equation $\dfrac{d^2z}{dt^2} + \omega_0^2 z = Z_0 \sin(\omega_1 t)$. Given that $\omega_0 \neq \omega_1$ find the solution for z given that $z = 0$ and $dz/dt = 0$ at $t = 0$.

8. The equation describing the small-angle oscillations of a simple pendulum is $\dfrac{d^2\theta}{dt^2} = -\dfrac{g}{l}\theta$, where θ is its angular displacement from the vertical at time t, l is the length of the pendulum and g is the acceleration due to gravity. Find an expression for θ as a function of t given that $\theta = \alpha$ and $\dfrac{d\theta}{dt} = \beta$ at $t = 0$.

9. A circuit consists of a capacitor C, a resistor R and a switch in series with a battery of emf V_0. The switch is initially open and the capacitor is uncharged. At $t = 0$ the switch is closed. The equation for the charge q on the capacitor as a function of time t after the switch is closed is $R\dfrac{dq}{dt} + \dfrac{q}{C} = V_0$. Find how the charge on the capacitor varies with time t given that $q = 0$ at $t = 0$.

10. A mass m on a spring is subjected to a damping force. The equation describing its displacement x from its equilibrium position as a function of time t is $m\dfrac{d^2x}{dt^2} = -kx - b\dfrac{dx}{dt}$, where $-kx$ is the force from the spring and $-bdx/dt$ is the force due to damping. The damping coefficient b is related to the spring constant k by $k = 4b^2/25m$. Find an expression for the subsequent motion of the mass given that $x = 0$ and $dx/dt = V$ at $t = 0$.

6.7   Graph Sketching – rational, hyperbolic and other functions

In addition to the concepts from previous sections, see
- Graph Interpreting - Level 6:
  isaacphysics.org/concepts/cm_graph_interpreting
- Graph Sketching - Level 6:
  isaacphysics.org/concepts/cm_graph_sketching

Toolkit #6. It is helpful to think about the following when interpreting a function $f(x)$.
1. What is the main shape of the function? (linear, quadratic, cubic, reciprocal, trigonometric etc.)
2. Where are the zeros i.e. the values of x for which $f(x) = 0$?
3. What happens to the function when $x = 0$?
4. What happens to the function as x gets very large?
5. Does the function diverge anywhere within its range.
6. What are the symmetry properties of the function?
7. Does the function have minima or maxima? If so, where?
8. Does the function have points of inflection?
9. Are there characteristic features of the function's gradient?

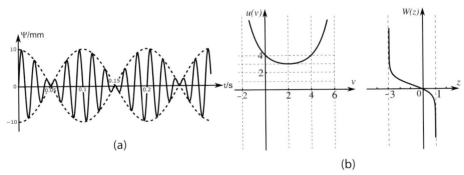

(a)

(b)

Figure 6.5

1. Consider two sinusoidal oscillations ψ_1 and ψ_2 of equal amplitude A such that $\psi_1 = A\sin(2\pi f_1 t)$ and $\psi_2 = A\sin(2\pi f_2 t)$, where f_1 and f_2 are their frequencies in Hz and t is the time in seconds. When combined they produce a signal $\Psi = \psi_1 + \psi_2$ as shown in Fig. 6.5a. Use the data in the graph to deduce the values of f_1, f_2 and A to 2 sf (assume $f_1 > f_2$).

2. The graphs of two functions $u(v)$ and $w(z)$ are shown in Fig. 6.5b; $u(v)$ and $w(z)$ are functions of cosh and inverse tanh respectively which have been translated and/or stretched and/or reflected. They are each defined by three positive constants: a, b and c for $u(v)$ and α, β and γ for $w(z)$. Use the information obtained from examining the graphs to deduce the forms of the functions and the values of the constants a, b and c and α, β and γ.

   a) Which of the following gives the correct forms for the two functions.
   - $u(v) = a\cosh(v + b) + c, w(z) = -\tanh^{-1}(\alpha(z - \beta)) + \gamma$
   - $u(v) = a\cosh(v - b) + c, w(z) = -\tanh^{-1}(\alpha(z + \beta)) + \gamma$
   - $u(v) = a\cosh(v - b) + c, w(z) = \tanh^{-1}(\alpha(z + \beta)) - \gamma$
   - $u(v) = a\cosh(v + b) + c, w(z) = \tanh^{-1}(\alpha(z - \beta)) - \gamma$
   - $u(v) = a\cosh(v - b) + c, w(z) = \tanh^{-1}(\alpha(z + \beta)) + \gamma$
   - $u(v) = a\cosh(v + b) + c, w(z) = \tanh^{-1}(\alpha(z - \beta)) + \gamma$
   - $u(v) = a\cosh(v + b) + c, w(z) = -\tanh^{-1}(\alpha(z - \beta)) - \gamma$
   - $u(v) = a\cosh(v - b) + c, w(z) = -\tanh^{-1}(\alpha(z + \beta)) - \gamma$

   b) Using the form of the function $u(v)$ you deduced in part a) find the values of the constants a, b and c to 2 sf.

   c) Using the form of the function $w(z)$ you deduced in part a) find the values of the constants α, β and γ to 2 sf.

3. Consider $r(s) = s(\ln s - 1)$ in the range $s \geq 1$. Answer the questions below about this function and hence sketch a graph of $r(s)$ for $s \geq 0$.

   a) Find the coordinates and nature of the stationary point of $r(s)$.

   b) Find the value of s at which $r(s)$ crosses the horizontal axis.

   c) Examine the behaviour of the gradient of the function $r(s)$ as $s \to 0$. By considering this, the sign of $r(s)$ as $s \to 0$ and the nature and value of the stationary point of $r(s)$, deduce which of the following statements about the behaviour of $r(s)$ as $s \to 0$ is correct.
   - $r(0) = 0; dr/ds = 0$ at $s = 0$ • $r(0) = 0$; as $s \to 0$, $dr/ds \to -\infty$.
   - As $s \to 0$, $r \to -\infty$ and $dr/ds \to 0$ • As $s \to 0$, $r \to -\infty$ and $dr/ds \to \infty$
   - $r(0) = 0$; as $s \to 0$, $dr/ds \to \infty$.   • As $s \to 0$, $r \to -\infty$ and $dr/ds \to -\infty$

4. A damped oscillator, forced to oscillate at a given frequency ω has an amplitude response $R(\omega)$ given by $R(\omega) = \dfrac{A}{\left((\omega^2 - \omega_0^2)^2 + \gamma^2\omega^2\right)^{\frac{1}{2}}}$ where A is proportional to the applied force, ω_0 is the frequency of free oscillations and γ is proportional to the amount of damping in the system. Answer the following questions about $R(\omega)$ for $\omega \geq 0$. Hence on the same axes, sketch the function $R(\omega)$ for $\omega \geq 0$ for $\gamma = \omega_0/10$ and $\gamma = \omega_0$.

a) Answer the **toolkit** questions 2 to 8 about the function $R(\omega)$.

b) Find the stationary points of the function $R(\omega)$. Give the coordinates of the maximum.

c) Give the coordinates of the other stationary point.

5. Consider the function $R(\omega) = \dfrac{Ab}{(\omega - \omega_0)^2 + b^2}$. (This is called a Lorentzian function or Lorentzian for short and has various applications in physics; for example the effect of collisions between molecules on the shape of the spectrum they produce can generally be described by a Lorentzian.) Answer the following questions about the function $R(\omega)$ for $\omega \geq 0$ and assuming $\omega_0 > b$. Hence, on the same axes, sketch the function $R(\omega)$ for $\omega \geq 0$ for $b = \omega_0/2$ and for $b = \omega_0/5$.

a) Answer the **toolkit** questions 2 to 8 about the function $R(\omega)$.

b) Find the value of ω at which the function has a maximum and the expression for R at this maximum.

c) Find the values of ω for which $R = R_{max}/2$ where R_{max} is the maximum value of R you found in part b). (The difference between these two values of ω is called the half-width of the function.)

d) Find the values of ω at which there are points of inflection.

6. Consider the function $r(s) = \dfrac{4 - s^2}{(4 - s)^2}$. Answer the questions below about this function and hence sketch a graph of $r(s)$. (You may find it helpful to sketch the two functions $(4 - s^2)$ and $1/(4 - s)^2$ and consider their product.)

a) Answer the **toolkit** questions 2 to 8 about the function $r(s)$.

b) Find the coordinates of the maximum of $r(s)$.

c) Find the value of s at the point of inflection.

7. Consider the function $f(q) = A\dfrac{\sin(qa/2)}{(qa/2)} = A\,\mathrm{sinc}(qa/2)$ where A and a are constants. Answer the questions below about this function and hence sketch a graph of $f(q)$ as a function of q in the range $\dfrac{-6\pi}{a} \le q \le \dfrac{6\pi}{a}$.

a) Find a general expression, in terms of a, π and n (where n is any integer except 0) for the values of q at which $f(q) = 0$.

b) Finding the value of $f(q) = A\dfrac{\sin(qa/2)}{(qa/2)}$ when $q = 0$ is not trivial as both the denominator and the numerator are equal to zero at $q = 0$. To deduce the value, write down the Maclaurin expansion for $\sin x$, divide this by x and then consider the limit of the resulting expression when $x \to 0$.

8. Consider the function $p(E) = \dfrac{1}{e^{(E-\mu)/kT} + 1}$, where $E \ge 0$ and μ, k and T are positive constants. Answer the questions below about this function, as-suming that $\mu > kT$, and then, on the same axes, sketch $p(E)$ as a function of E for $T = 0.001\mu/k$ and for $T = 0.1\mu/k$. On a separate graph sketch the function $n(E) = AE^{\frac{1}{2}}p(E)$, where A is a constant, for $T = 0.1\mu/k$. (You do not need to know what these functions represent in order to answer this question. However, if you are interested, $p(E)$ is called the Fermi-Dirac dis-tribution and applies to a system of fermions (particles with half-integral spin such as electrons and protons) in thermal equilibrium at temperature T; it gives the probability of finding a fermion in a state of energy E. The constant μ is called the chemical potential (or Fermi energy) and k is the Boltzmann constant. In a simple model of a metal $n(E)$ then represents the number of states of energy E which are occupied at temperature T.)

a) Find an expression for $p(\mu)$.

b) Find an exact expression for $p(\mu - kT)$.

c) Find an exact expression for $p(\mu + kT)$.

d) Find the value of $p(0)$ when $T = 0.1\mu/k$, giving your answer to 4 sf.

e) Hence, without further calculation, deduce the value of $p(0)$ when $T = 0.001\mu/k$, giving your answer to 4 sf.

f) Deduce the limiting value of $p(E)$ as $E \to \infty$.

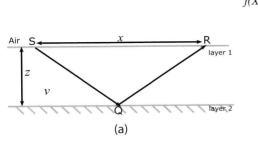

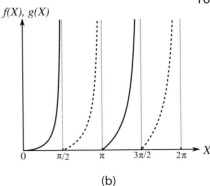

(a)

(b)

Figure 6.6

9. Consider a geophone placed on a sedimentary rock (at R) and at a certain distance away a seismic source (at S) generating P-waves (e.g. a hammer hitting the ground). As for light, these seismic waves will reflect and refract at interfaces where the speed of the wave changes. Assuming a two-layered model of the subsurface as shown in Fig. 6.6a, the time taken for the reflected wave SQR to reach the receiver R located a distance $x \geq 0$ from the source S is given by $t_r(x) = \dfrac{2}{v}\sqrt{z^2 + \dfrac{x^2}{4}}$ where v is the speed of the seismic wave in the upper layer (layer 1) and z is its thickness; this is the equation of a hyperbola. The travel time for the direct wave (travelling through layer 1 from S to R in a straight line) is given by $t_d(x) = \dfrac{x}{v}$. Answer the questions below about $t_r(x)$. Hence, on the same axes sketch $t_r(x)$ and $t_d(x)$.

a) Answer the **toolkit** questions 2 to 8 about the function $t_r(x)$.

b) Find the value of x at which the function $t_r(x)$ has a minimum and give the corresponding expression for the minimum value of $t_r(x)$.

c) Give the expression for $\dfrac{dt_r}{dx}$ when x gets very large and positive.

d) On the same axes, sketch $t_r(x)$ and $t_d(x)$. Which of the following statements about the relationship between $t_r(x)$ and $t_d(x)$ is true?

- At large distances, the travel time curve for the reflected wave is asymptotic to the travel time curve for the direct wave.
- The travel time curve for the reflected wave intercepts the travel time curve for the direct wave.
- The travel time curve for the reflected wave tends to infinity much faster than the travel time curve for the direct wave.

10. A graphical method is required to find the quantised energy levels of a particle of mass m in a potential well of depth V_0 and width a. By considering the solutions of the Schrödinger equation inside and outside the well we define two positive variables X and Y; there are two classes of solutions which arise which require that $Y = f(X)$ or $Y = g(X)$, whilst at the same time $X^2 + Y^2 = R^2$ (where R is related to V_0, a and m). The functions $Y = f(X)$ (solid lines) and $Y = g(X)$ (dashed lines) are shown in Fig. 6.6b.

The energy levels are given by

$$E_n = -\frac{Y_n^2}{X_n^2 + Y_n^2} V_0$$

where (X_n, Y_n) are the coordinates of the points of intersection of $f(X)$ and $g(X)$ with $X^2 + Y^2 = R^2$.

a) Which one of the following statements about $f(X)$ and $g(X)$ is correct?

- $f(X) = \tan X$, $g(X) = -\cot X$ • $f(X) = \tan X$, $g(X) = \cot X$
- $f(X) = X \tan X$, $g(X) = X \cot X$ • $f(X) = \tan X$, $g(X) = -X \cot X$
- $f(X) = X \tan X$, $g(X) = -X \cot X$ • $f(X) = X \tan X$, $g(X) = \cot X$

Consider the point/points of intersection of the function $X^2 + Y^2 = R^2$ with the functions $f(X)$ and $g(X)$ shown in Fig. 6.6b and answer the following questions.

b) Deduce the condition on R for which there will be one and only one solution for this set of equations.

c) In the range $a < R \leq b$ there are three and only three solutions for this set of equations; deduce the values of a and b.

d) Deduce the condition on R for which there will be more than N solutions.

e) The function $X^2 + Y^2 = 16$ intersects the functions $f(X)$ and $g(X)$ at $(1.25, 3.80)$, $(2.48, 3.14)$ and $(3.60, 1.75)$. In this case the highest value for the energy is given by $E = \alpha V_0$. Deduce the value of α to 2 sf.

Applications to Sciences

> We look at vectors problems that may be unfamiliar: Random vectors can be added to give random walks. They permeate nature from the very small scale, for instance nuclear spins, to the molecular (polymer molecules), the microscopic (the dancing of pollen grains – Brownian motion), to the galactic. Techniques from the preceding sections will be employed.

1. The time shown on a clock changes from 4:00 to 4:30. The minute hand, of length 25 cm, moves smoothly halfway around the face. The movement of the tip of the minute hand can be thought of as lots of small displacement vectors taking the tip from the old to the new position. Where vector answers are required below, give them in terms of unit vectors $\hat{i}$ and $\hat{j}$ pointing from 12 o'clock to 6 o'clock and from 9 o'clock to 3 o'clock respectively.

   For the tip of the minute hand, what is the

   a) sum of the displacements of the tip during this half hour?

   b) total distance travelled by the tip?

   c) average speed of the tip? (Use cm/minute for speeds.)

   d) instantaneous speed of the tip?

   e) instantaneous velocity at 4:15?

   f) instantaneous velocity at 4:10?

   g) average velocity during the half hour?

   h) average acceleration during the half hour? (Use units of cm/minute^2.)

   i) average displacement of the tip, over the half hour, from its position at 4:00?

2. Plane waves from the sea travel at speed u and meet a beach obliquely – wavefronts arrive with their normals at an angle θ from the normal of the

beach. As a result, the point of contact of the arriving wave moves along the beach at speed v. Calculate v.

For what angle of incidence measured from the normal does v exceed the speed of light, c? Is this a meaningful question?

Random Walks

An air molecule moves randomly, colliding with other moving air molecules, changing direction at random, and travelling in a straight line until the next collision. This 'Random Walk' motion can be modelled with simple mathematics which, along with experiment, yields microscopic quantities associated with gases – how many collisions a molecule makes each second, and how far it travels between collisions.

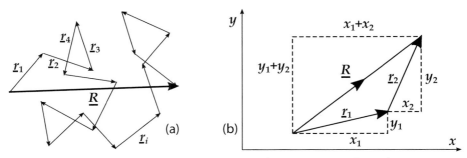

Figure 7.1. (a) N random steps shown as vectors $r_i = (x_i, y_i)$ adding to form a resultant vector R. (b) Two individual vectors showing their components adding together to give the components of $r_1 + r_2$.

A particle moving randomly in this way is unlikely to end up back at the starting point, but is also unlikely to travel in a straight line and end up at the maximum possible displacement. It is most likely to end up somewhere between these two distances, at a displacement R from its starting point (see Fig. 7.1a). This process is *diffusion*.

What can we find out about this average displacement R?

Analysis of 2-steps in 2-D

$$R = r_1 + r_2$$

The direction of R will be random. But what can we deduce about its length

R? Resolving $\mathbf{R}$ into components (see Fig. 7.1b) gives

$$R_x = x_1 + x_2 \quad \text{and} \quad R_y = y_1 + y_2$$

and using Pythagoras' theorem:

$$
\begin{aligned}
R^2 &= (x_1 + x_2)^2 + (y_1 + y_2)^2 \\
&= x_1^2 + x_2^2 + 2x_1x_2 + y_1^2 + y_2^2 + 2y_1y_2 \\
&= [x_1^2 + y_1^2] + [x_2^2 + y_2^2] + 2x_1x_2 + 2y_1y_2 \\
&= r_1^2 + r_2^2 + 2x_1x_2 + 2y_1y_2
\end{aligned}
$$

Analysis of N steps in 2-D
Extending this analysis to N steps gives

$$
\begin{aligned}
R^2 &= [r_1^2 + r_2^2 + r_3^2 + ... + r_N^2] + [2x_1x_2 + 2x_1x_3 + 2x_1x_4 + ...] \\
&+ [2y_1y_2 + 2y_1y_3 + 2y_1y_4 + ...]
\end{aligned}
$$

The terms like x_1x_2, x_2x_3, x_2x_4 etc. will have positive and negatives values at random and so will cancel out when averaging over steps, denoted by $\langle ... \rangle$. Therefore

$$\langle R^2 \rangle = \langle r_1^2 + r_2^2 + r_3^2 + ... + r_N^2 \rangle$$

The N steps have the same average length, a, where $a = \sqrt{\langle r_i^2 \rangle}$. Therefore

$$\langle R^2 \rangle = Na^2 \rightarrow \sqrt{\langle R^2 \rangle} \equiv R_{rms} = \sqrt{N}a,$$

where R_{rms} denotes "root mean square". In summary, if a molecule takes N steps in random directions, then *on average* its end point is a distance $\sqrt{N}$ steps away from its starting point. Compare this to the *total length* of the path travelled, $L = Na$, which is much greater than R (by a factor of $\sqrt{N}$) if N is large.

3. On average what will be the magnitude of the displacement, in terms of a, of a molecule undergoing a random walk if it makes
   a) 16 steps?   b) 10 000 steps?   c) 10^6 steps?

d) If the average molecular speed is v, what is the mean number of steps taken in a time t?

e) Using part d), how does the rms distance R_{rms} vary with time?

f) If a poisonous gas was to be released into the room, would we be more concerned about diffusion or convection in the room as a means of spreading the gas?

4. Bromine diffuses a distance of 20 cm in still air in 500 s. The bromine molecule has a mass five times that of an air molecule. By randomising energy through collisions, all molecules in the gas have the same average energy. Thus Br molecules will travel on average $1/\sqrt{5}$ the RMS speed of air molecules. [Prove this must be so.]

a) If the average speed of an air molecule is about 500 m s^{-1}, estimate the average distance between collisions of a bromine molecule (this is known as the mean free path).

b) How many collisions does a bromine molecule make each second?

c) Since the particles are moving in random directions why should the brown bromine front be seen to move through the clear air?

d) *If the gas mixture is at atmospheric pressure (10^5 Pa) and at temperature 300 K, what is the radius of a bromine molecule, taking it as a sphere? [Hint; use the perfect gas law to find the number of molecules per unit volume. Assume molecules move through their share of the total volume before, on average, colliding.]

5. Cosmic rays are charged particles that move randomly in a galaxy as a result of being scattered by randomly oriented interstellar magnetic fields. This process resembles diffusion. In such situations we can apply a random walk to the cosmic rays reaching the Earth from beyond our galaxy (the Milky Way). Given the dimension of our galaxy as 5×10^{20} m and if the mean free path for a cosmic ray is 3×10^{18} m, estimate how long it takes for a cosmic ray to travel across the galaxy. Assume that it travels at almost the speed of light.

7.2   Advanced vectors – 2

Two initially less familiar vectors are area, A, and flux (density), S. Both are of huge importance throughout physics. Diagrams are vital, as usual.

Area

Vector area: Figure 7.2a shows a plane, with some vectors u, v, w in the plane, and vector A drawn perpendicular to the plane. A is unique to the plane, indicates the direction of the plane and has magnitude equal to its area.

Vectors u, v, w are all perpendicular to A. Express this property of the vectors in the plane using the notion of the dot product (see concept page Vectors - Resolving Vectors: Level5 – Dot (or Scalar) Product at isaacphysics.org/concepts/cm_vectors2).

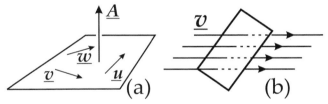

Figure 7.2. (a) The area vector A of a plane. (b) Flux of a fluid with velocity v passing through a surface.

1. Draw a cube of edge a sitting on a table.

   a) What magnitude is each face's area vector?

   b) Give the Cartesian forms (x, y, z) of the area vectors, where x, y, z are their projections along axes aligned with the cube and centred in one of its corners. Draw the axes in your diagram. (Although vectors are not generally fixed to any point in space, it is useful to think of area vectors emerging perpendicularly from their associated surface.)

   c) What relation does the sum, S, of the vector areas of the faces not in contact with the table have to the vector area, S', of the face on the table?

   d) What is the magnitude of the total vector area of the closed cube?

Flux

The word "flux" comes from the Latin "fluxus" meaning "flow", and is used in physics to describe something that physically or conceptually flows through space. For example, the flux of particles through space, the flux of energy from the sun incident on a solar panel, or the magnetic flux through a wire loop; see Fig. 7.2b. Flux can either be expressed as a rate of flow *per* unit cross sectional area (an *intensive* quantity), or it can be calculated for a given area (an *extensive* quantity). Strictly speaking, the former should be called "flux density", but the term "flux" is sometimes used in physics in both senses.

The units of flux density depend on the context in which it is used. For example, the flux density of rain drops has units (number of droplets) $m^{-2}s^{-1}$, but the flux density of the mass of water has units kg $m^{-2}s^{-1}$.

It is important to note that, within electromagnetism, flux density refers to the Magnetic Field Strength B with unit tesla, T. By contrast, the total magnetic flux through a given area A is the flux $\Phi = BA$ (if B falls normally on the loop: that is, if it arrives along the A direction). Here, flux is an extensive quantity, with unit the weber, W. But throughout this section, flux density is used, and is an intensive quantity.

2. a) Show that the units of the flux density of people entering a football stadium are $m^{-1}s^{-1}$. Estimate its value at an entrance gate at peak entry time.

b) Answer the equivalent question for cars on a *one-way* road. Estimate the peak flux density on one side of a motor way. Discuss the factors that govern the flow?

c) What is the average flux density for the motor way considering both sides?

Flux density dotted with area – capture

How much material (or energy or mass or number of field lines, etc.) is intercepted per second by a surface from an incident flux density of these quantities? The magnitude of the surface is clearly important. Flux density is per m^2 and the larger the surface, the more is captured. If the flux density is parallel to the surface, nothing is captured. It is the component of S perpendicular to the surface that is important, that is the projection of S along

A. Taking into account the sizes of both *S* and *A*, the capture rate is therefore $S \cdot A$, where $\cdot$ denotes the dot product (scalar product) of *S* with *A*, that is $SA \cos \theta$ with θ the angle between the vectors. For instance, when flux density is parallel to the surface, *S* is perpendicular to *A* and hence the dot product is zero.

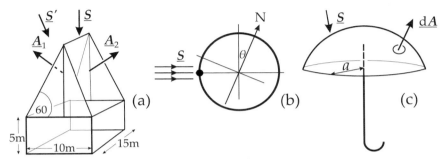

Figure 7.3. (a) A house with a sharply pitched roof with vector areas A_1 and A_2, with two examples of rain flux density *S* and *S'*. (b) The Poynting vector *S* of radiation from the Sun falling on the Earth. The Earth's axis of rotation is tilted by θ away from the orbital plane's normal. (c) A sun umbrella in the form of a spherical cap with circular perimeter of radius *a*. A small element of area d*A* is indicated, along with the solar flux density *S*.

3. Draw a square wire loop of area *A* rotating with angular speed ω about an axis that is perpendicular to the flux density *B* and in the plane of the loop.

   a) If the loop normal is initially along *B*, give the rate of change of the captured flux $\Phi(t)$.

   b) What current $I(t)$ flows in the loop if it has a resistance *R*? [Look up Faraday's Law.]

4. During a heavy rain storm, take the mass flux density of falling water to be of magnitude $S_m = 3 \times 10^{-3} \text{kg m}^{-2} \text{s}^{-1}$.

   a) Find the magnitude of this volume flux density S_v in SI units.

   b) What is the value in (old fashioned – inch/hour) units sometimes used by the weather service in the UK?

c) Rain falls vertically onto a house with footprint 10m×15m and a steep A-frame style roof pitched at 60°; see Fig. 7.3a. The magnitude of the rain flux density is S_m. Considering the two roof areas A_1 and A_2, find the volume of water per second that can be collected from the guttering, expressing your answer in litres per second.

d) How long does it take to collect a tonne of water?

e) At what rate would the water accumulate if the roof blew off and it rained directly on to the floor? Comment.

f) A strong wind blows from the side of the house with the A_1 roof so that the rain drives in with S' inclined at 30° to the vertical (where $|S'| = |S|$). At what rate is the rain now collected by the roof? (Assume the rain is propelled onto the house, ignoring the airflow being deflected close to the house to go around it.)

g) How much is intercepted by the house as a whole?

5. It is noon, with the Sun overhead, at the particular location indicated by the dot in Fig. 7.3b.

a) What days of the year could it be?

b) What is the latitude there?

c) What large city is approximately in this position?

d) At what southerly or northerly latitude is there exactly one day a year where the Sun never rises?

6. A black tent has the shape of a cube of edge a.

a) When the sun is directly overhead, how much power is being absorbed by the tent if $a = 2.0$m? Compare this power with that of a 1 kW electric heater. [The Sun's intensity of energy flow when overhead on Earth is 1.4 kWm^{-2}, the Sun's energy flux density at the Earth's orbital radius.]

b) The Sun is now 45° above the horizon and shines only onto one side and onto the top. How much power is now being absorbed?

c) The tent is lifted up to leave only the (black) ground sheet. How much power is the ground sheet absorbing with the Sun still at 45°?

d) The Sun is now shining along the line of a body diagonal (the vector con-necting two opposing corners of the cube). Draw in the diagonal on your cube. How long is the body diagonal? Form the unit vector of the diagonal, using a suitable coordinate system based on the cube itself. Calculate what power is now being absorbed by the tent.

7. a) What is the resultant vector area of the curved surface of the umbrella of Fig. 7.3c? [Hints: Return to question 7.2.1c) which illustrates a general result – the sum of the area vectors of a closed surface is zero. Can you now prove it? Consider also the current question, taking an incident flux to pass right through the umbrella.]

b) Now take the umbrella as a hemi-spherical shell of radius R, and the solar flux density to be S: What power hits the umbrella if the Sun is overhead?

c) What power arrives if the axis is held perpendicular to S?

d) *The Sun is at $38°$ to the horizon. How much power does the umbrella receive when held vertical?

e) General knowledge discussion: What day of the year is it in part d) if it is noon on a spring day in Cambridge?

7.3   The calculus of change – Exponentials

Exponentials were first noticed in the 16th century in connection with compound interest on money. A fascinating book about exponentials, logarithms, their discovery, their mathematics and their applications is ' "e": The Story of a Number' (Princeton Science Library) by Eli Maor. Exponential variation, whether in time or space, is a particular and ubiquitous form of change, the rate of which is linearly proportional to the quantity or number currently present.

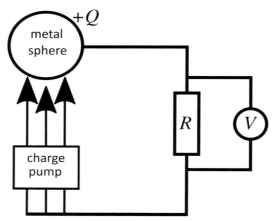

Figure 7.4. A metal sphere charged up with positive charge and discharging through a resistor

1. Discharging a charged metal sphere
   A metal sphere/dome is charged with positive charge using a high voltage generator; see Fig. 7.4. The dome is continuously charged such that a constant current flows through a resistor $R = 1.0 \times 10^{11}\ \Omega$. The potential shown on the voltmeter is 150000 V .

   a) Calculate the current I flowing through the resistor.

   b) The generator is now stopped at time $t = 0$, so that the charge on the dome starts to leak away: What is the initial rate at which the charge leaks away through the resistor?

c) The potential V on the dome is proportional to the charge on the dome, with constant of proportionality C (the capacitance), that is $Q = VC$. What is the charge, Q_0 at time $t = 0$?

d) Relate I to V and to rates of change of Q to derive an expression for dQ/dt.

e) Recognise the form of the solution to this equation is $f(t/\tau)$, and give the time constant τ characterising changes in charge remaining on the dome.

f) What is the solution $Q(t)$, given $Q = Q_0$ at $t = 0$.

2. People

Without intervention, the rate of population growth is proportional to the population, P, at the time. Malthus first recognised that this dependence must lead to disaster(s), the nature of which he, and others since, have speculated on. This is known as the Malthusian catastrophe. What are examples of these disasters?

a) What form does $P(t)$ take if the *rate* of growth is $r\%$ per year?

b) In the USA currently the growth rate $r = 1.5\%$ per year. What is the population doubling time? (Give your answer in years.)

c) How can Malthusian catastrophes be avoided?

3. (Isaac) Newton's Law of Cooling

Newton's Law of Cooling states that the rate of cooling of a hot body is linearly proportional to the temperature difference between that of the body, T, and that of its surroundings, T_S: the greater the difference, $T - T_S$, the faster the cooling.

a) Take I for the constant of proportionality. What are its units?

b) Write down Newton's law for the rate of change of temperature with time.

c) Show that the result above can be written as $\dfrac{d\Theta}{dt} = -I\Theta$ and give the definition of the new variable Θ.

d) Solve the above equation and give an expression for the time constant of the cooling.

e) At what time is the temperature difference with the surrounding half the initial value?

4. Money!

A sum of money **m** is invested and earns interest at a rate p. [Note: p is a *rate*, that is "per unit of time".]

a) If $p = 10\%$ per year, and the interest is calculated and compounded annually, by what multiple has your money increased after 10 years? ["Compounded" means the interest is added to your capital, whereupon interest is paid on the enhanced sum.]

b) For continuously compounded interest at a rate p %, the sum of money m increases by $dm = (mp/100)\, dt$ in time dt. Check the units of the previous expression – reassure yourself they are OK.

c) Find an expression for $m(t)$ if you start with m_0 at time $t = 0$.

d) How much is your money worth after 10 years of being continuously compounded at rate $p = 10\%$ per year? Compare your answer to that in part a).

e) If m is the capital of a charity, payments are made at a rate q, and interest is continuously compounded at a rate of p %, show that the equation for m is now

$$\frac{dm}{dt} = \frac{p}{100}m - q$$

If the Charity Commissioners demand that capital does not decrease, what is the maximum rate of payments allowed?

Money is revisited in Exercise 7.3.9.

5. Nuclear chain reactions

Neutrons are released by fissile nuclei spontaneously decaying, for instance ^{235}U with a half-life of 700 million years. If these neutrons are slowed down (moderated) by collisions with the nuclei of other light elements such as deuterons, then they can be subsequently absorbed by other fissile nuclei and induce them in turn to decay, producing a number d of daughter neutrons, plus two or more fission products (lighter elements, for instance barium, krypton and others; $d = 3$ for this example of ^{235}U); see Fig. 7.5.

If neutrons moderate at a rate r, then the increase in the number density n of neutrons in time Δt is

$$\Delta n = (d - 1)\, n\, r\, \Delta t$$

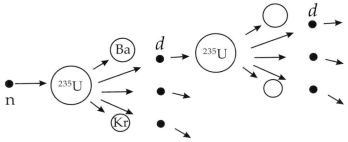

Figure 7.5. A neutron n approaches a fissile uranium-235 nucleus causing it to fission into two daughter nuclei (here barium and krypton for example) and d daughter neutrons. The path of one of these is shown — it moderates, is absorbed in turn in another uranium, and causes a net increase of $(d - 1)$ in the neutron population.

For each neutron absorbed there are effectively $(d - 1)$ extra ones produced if d are then emitted after the absorption. The change in the number of neutrons is proportional to the number n already present. Each of these n neutrons finds a nucleus at a rate r (number of decays per unit of time) and hence $n\,r\Delta t$ neutrons find a nucleus in a short time Δt. Therefore, $dn/dt = (d - 1)r\,n$.

a) Derive the differential equation $dn/dt = (d - 1)r\,n$. The solution, $n(t)$, will be in the form $f(t/\tau)$. Give the time constant, τ, characterising the changing neutron population.

b) How does the number density of neutrons $n(t)$ change with time from an initial background level of n_0 at time $t = 0$?

c) If the value of d is 1.002 for a nuclear reactor, and the average time between fission events is 12 ms, calculate the time required for the reactor power (proportional to neutron number) to double.

d) If the value of d is 3, as in a nuclear weapon, and the average time between fission events is reduced to 2 ms, calculate the time required for the fission power to double.

e) Neutrons are also lost at a rate r_1 due to absorption by non-splitting nuclei, either inert isotopes of heavy elements, or by protons. Revisit the above argument for increase and amend it for this cause of decrease to give a new equation for dn/dt.

f) How does $n(t)$ now vary with time?

g) What is the condition for a self-sustaining chain reaction?

h) If the reaction is supercritical, after what time t_{100} is the number density of neutrons 100 times greater than background?

6. A leaky tank

A tank with a leak empties in a way entirely analogous to a capacitor being discharged, Exercise 7.3.1.

A cylindrical tank of cross-sectional area A, filled with liquid to an initial height of h_0, has a small hole in its base. Fluid leaks out at a volume flow rate I where $I = ap$, with p the pressure difference between the fluid at the bottom of the tank and the outside, and a is a constant (discuss how this constant might depend on the hole). The hydrostatic pressure difference depends on the depth of the fluid: $p = \rho g h$, where ρ is the fluid density, g the acceleration due to gravity, and $h(t)$ the current height of liquid in the tank at time t.

a) Find how $h(t)$ varies as the tank empties.

b) Fluid is replaced at a rate of r. What is the equilibrium height h_{eq} of fluid in the tank at long times?

c) If the tank is empty ($h_0 = 0$) at time $t = 0$, find how $h(t)$ increases to the value h_{eq} over time.

d) What is the time constant (the characteristic time) for this process?

7. Discharging a capacitor

A capacitor stores opposite charges on its two plates, the potential difference V between the plates being proportional to the charge stored:

$$V = \frac{1}{C}Q$$

where the constant of proportionality is the inverse of the capacitance, C. Charge flows in and out at a rate depending on how much already resides on the capacitor – a classic recipe for an exponential! See the circuit of Fig. 7.6.

a) Will the capacitor charge at a constant rate?

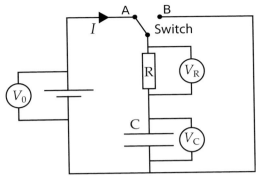

Figure 7.6. When the double throw switch is connected to A, the capacitor is connected through a resistor to a cell of potential difference V_0. When the switch is connected to B, the capacitor is discharged through the resistor.

b) Give an expression for Q_0, the magnitude of charges on the plates when charging is complete. Note that the *net* charge on the capacitor as a whole is zero.

c) When the capacitor is charged, energy is stored since the charges +Q and -Q are separated. Why does separation imply stored energy.

d) The switch is now moved to position B so that the battery is disconnected from the capacitor. Charge now flows through the discharging circuit at a rate I, reducing the stored charges $\pm Q$, and the potential difference across the capacitor decreases with time from V_0 to $V(t) = \frac{1}{C}Q(t)$. Sketch on the circuit the direction of current flow.
Find an equation for dQ/dt. Care with signs!

e) Write down, by inspection, the solution to this equation with and appropriate initial condition.

f) What is the time constant in this discharge?

g) Write an expression for the the variation with time of potential difference $V(t)$ across the resistor?

8. A bouncing ball
   The height of a bouncing ball can be expressed as a geometric progression. Quantities changing geometrically are in fact exponential processes: they change in direct proportion to themselves. Algebraic manipulation exposes

the natural exponential form.

A ball is dropped from an initial height h_0 on to a table such that the height of the bounce is $h_1 = \alpha h_0$. Draw a diagram showing several bounces.

a) Using the properties of powers, exponentials and logarithms, show that the height of the nth bounce as a function of n is $h_n = h_0 e^{cn}$ and give an expression for the constant c. Reassure yourself that this is a decaying function.

b) How far does the ball travel before it stops bouncing, in the ideal case?

c) Why might this distance not be attained in practice?

d) Find the time between the nth and $(n + 1)^{\text{th}}$ bounces. Show that it can be written as $2t_0 e^{-\beta n}$ and give expressions for t_0 and β.

e) Find the total time spent bouncing in the ideal case. How many bounces does the ball execute in this time?

f) * Calculate the frequency of bouncing heard as a function of bounce number n and also of time T_n at the nth bounce.

g) Discuss any divergences that arise!

9. Money revisited

A sum of money m is invested with an interest rate of p %. Banks can choose how often throughout the year to calculate the interest and whether or not to compound the interest. For example, interest might be calculated annually, quarterly, monthly, or continuously.

a) If $p = 10\%$ per year, by what factor does your capital increase after 10 years if the interest is compounded quarterly?

b) By what factor does it increase after 10 years if the interest is compounded monthly?

c) Give an algebraic expression for the factor by which your money is worth more after t years when the interest is compounded n times a year.

d) When money is compounded continuously, the interest is effectively being applied an infinite number of times throughout the year.

The exponential function can be defined as follows:

$$e^x = \lim_{n \to \infty} (1 + \frac{x}{n})^n$$

Using this definition and comparing it with your answer above for compounding n times per year, show that when interest at rate p% per year is continuously paid and compounded, your capital increases exponentially. Give the exponent of the increase factor after t years.

Note that this is the same expression as that derived by derivatives in Exercise 7.3.4 b).

e) To enable consumers to compare interest rates which have different compounding frequencies banks often publish the AER (annual equivalent rate). This gives the equivalent interest rate as if it were applied only once a year.

Would you prefer your interest of p % per year to be compounded annually, quarterly, monthly or continuously? Calculate the AER for quarterly, monthly and continuously compounded interest. [Extension: if you have access to a computer, plot graphs to show the increase in your capital for yearly, quarterly, monthly and continuously compounding interest at various choices of rates p % per year.]

f) If m is the capital of a charity, payments are being made at a rate q per year, and interest is accruing at a rate of p % per year, what is $m(t)$ after t years if the capital is initially m_0? This follows on from Exercise 7.3.4 e).

7.4   Words to Physics to Calculus

Turning words into physics and into differential equations
The most difficult part of physics and maths is reading a problem as
posed, understanding the essential physics, and then converting it into
a framework (physics analysis and mathematics) where it can be solved.
The first pass through the questions practises these steps. Then you are
invited to actually solve the mathematics and comment on the physics
that is thereby revealed.

Reading and writing!

Closely read the text of the following problems, and then convert your understanding into symbolic form. Drawing a (large) diagram as you proceed is
a great aid to both reading and understanding. The marginal diagram here
is along the right lines, but it should be large – see page 127.
The first question appears twice. In the second version the important words
and notions are underlined, and the meaning a scientist extracts from them
appears in blue text. You should be reading and underlining in subsequent
questions, making a mental note of the implications of the words.

1. A particle of effective mass m sediments through a fluid, receiving a retard-
    ing force proportional to its velocity. Denoting its speed by v, and the rel-
    evant constant of proportionality by k, write an equation that describes the
    motion.

    A particle of <u>effective mass</u> m

        mass suggests Newton's Second Law and inertia, and also suggests grav-
        ity, weight force etc.; effective suggests there may be other effects act-
        ing, but that you can incorporate them into an effective m – here the
        other effect is an Archimedian upthrust

    <u>sediments</u>

        suggests falling under gravity, but slowly (and thus a drag proportional
        to speed – but you will be given that here)

    through a fluid, receiving a <u>retarding</u>

        means oppositely directed to the velocity, i.e. reducing the downward
        acceleration

force
> suggests Newton's Second Law, equations of motion

proportional to its velocity
> force directed along the motion, and clearly "retarding" means the pro-
> portionality is negative

a) Draw a diagram.

b) Write down a differential equation for the velocity of the particle.

2. A mass m lies on a smooth horizontal surface and is attached to one end of a massless spring of spring constant k, the other end of which is anchored. The displacement of this mass, away from where the spring takes its natural length is x, which can take positive and negative values.

a) Draw a diagram showing the forces acting on the mass. Ignore gravity.

b) Write down the differential equation for the displacement x of the mass, that is an equation involving x and its derivatives with respect to time.

c) The mass now suffers a retarding force proportional to its velocity (with constant of proportionality q). Write down the modified equation of motion.

3. A ball of mass m is dropped and moves quickly through air, receiving a drag force proportional to the *square* of its speed.

a) Draw a diagram showing the forces acting on the ball.

b) Write down an equation that describes its *velocity* as a function of time. Denote the appropriate constant of proportionality by q.

4. * A rocket, currently of mass m, burns fuel at a steady rate α (mass per unit time). The fuel leaves with a speed v_0 *relative to the rocket*. Write down an equation for the acceleration of the rocket as seen by a stationary observer. Neglect gravity, drag, steering etc.

Solve the problems

Now complete each of the above problems, which is the simpler task of solving the equations that you have formulated. Refer on-line to the Isaac concepts for differential equations, and to sections 5.5 and 6.6.

Problems need posing correctly: if reducible to a first order differential equation (one order of derivative), they need one "boundary" or "initial" condition; second order differential equations require two conditions, and so on.

5. A particle of mass m sediments through a fluid, receiving a retarding force proportional to its velocity. It starts from rest, i.e. its speed has $v(t = 0) = 0$. Give the variation of speed with time during the motion. Also give the variation of position x with time, given that $x(t = 0) = 0$. [Note that another condition has crept in. Why?]

6. A mass m is attached to one end of a massless spring of spring constant k, the other end of which is anchored. The displacement of m, away from where the spring takes its natural length, is x. Ignore drag forces and gravity.

   a) Find $x(t)$ if at $t = 0$ one has $x = A$ with the particle at rest. Interpret A.

   b) Find $x(t)$ if at $t = 0$ one has $x = 0$ and the particle moves with speed v_0. Relate v_0 and A.

7. A ball of mass m is dropped and moves quickly through air, receiving a drag force proportional to the square of its speed, v, (with constant of proportionality q). Write down the equation of motion in terms of v.

   a) What is the terminal speed, v_f say?

   b) *If $v(t = 0) = 0$, describe how the speed subsequently varies.

   c) *Find how the distance dropped through, x, varies with time.

8. A rocket, currently of mass m, burns fuel at a steady rate α (mass per unit time). The fuel leaves with a speed v_0 *relative to the rocket*. Write down an equation for the acceleration of the rocket as seen by a stationary observer. Neglect gravity, drag, steering etc. If the initial mass is m_0 and the rocket is initially at rest, give the final speed v_f when the mass is then m_f.

9. A ball of mass m is propelled upwards, moving quickly and thus receiving a drag force proportional to the square of its speed (with a constant of proportionality q).

   a) Write down the equation of motion in terms of v and its derivative.

   b) In order to solve the equation in part a) it is useful to simplify it by defining a characteristic time $T = \sqrt{m/gq}$ and a characteristic speed $V = \sqrt{mg/q}$.

      i. Rewrite the differential equation in terms of reduced speed $u = v/V$ and reduced time $\phi = t/T$.

ii. Notice now how simple the equation now appears. Discuss possible advantages of this change (see online discussion).

c) Solve the equation in part b) for $u(\phi)$. Given that the initial upward speed is $v(t = 0) = v_0$, calculate the speed $v(t) = u(\phi)V$ describing the upward phase of the ball's motion. Give your answer in terms of V and T and v_0.

d) Give the reduced time t_f/T when the ball comes to rest.

e) Compare your answer in part b i) with question 7 b).

f) Find how the distanced travelled upwards, x, varies with time.

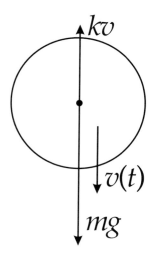

7.5   The calculus of change – Population

> **More complex change.**
> Most phenomena involve change, with respect to time (dynamics, pop-
> ulation, money etc.) or space (waves, diffusion, . . .) or change in other
> controlling variables. Changing phenomena are described by (differ-
> ential) equations involving rates of change (derivatives in calculus). We
> have seen a wide range of systems that involve exponential change, but
> change can be more complex.

Population growth and its regulation

The number of water weed plants at time t in a pond of fixed size is $f(t)$.
This population increases at a rate proportional to (i) the number of plants
already existing, and (ii) the rate, a, of supply of nutrient to the pond water
(e.g. from sediment). Identify these effects in the rate equation for f:

$$\frac{df}{dt} = af \qquad\qquad (7.1)$$

1. a) What is the fractional increase in population with time, $f(t)/f_0$, where f_0
   is the initial population.

   b) What is the characteristic time of the population increase?[1]
   See section 7.3. Malthus showed such behaviour leads to catastrophes.

If each plant actually consumes nutrients at a rate a_1, then $a_1 f$ is removed
from the flow of nutrient to the general population of plants, giving an effective rate $a - a_1 f$ instead of a. Putting this nutrient flow into the above
population equation, the rate of increase of population is reduced to

$$\frac{df}{dt} = (a - a_1 f)f. \qquad\qquad (7.2)$$

The equation can be made to appear simpler by "reducing" time and population: f is a number (no dimensions), hence the left hand side has dimen-

---

[1] See question 7.3.1 e) for casting equations into a form yielding a characteristic time for
a process.

sions of inverse time, as must the right hand side, that is a and a_1 which then both have the dimensions of inverse time. Divide both sides by a, and, on the left-hand side, put a together with the t. The combination $at = u$ is dimensionless and measures time in units of $1/a$. For this problem:

$$\frac{df}{adt} = \frac{df}{du} = (1 - \frac{a_1}{a}f)f.$$

Now multiply both sides by $\frac{a_1}{a}$ and put this factor together with f. Call $z(u) = \frac{a_1}{a}f(u)$, which is the number of plants measured relative to the quantity a/a_1, which sets a scale for plant numbers for this system. Then

$$\frac{a_1}{a}\frac{df}{du} = (1 - z)\frac{a_1}{a}f \rightarrow \frac{dz}{du} = z(1 - z), \qquad (7.3)$$

which is the Logistic differential equation (Verhulst, 1844) for $z(u)$. It *looks* very simple, without any parameters. This is because we have measured both time and plant number in terms of their characteristic values, $1/a$ and a/a_1 respectively. In physics such equations occur in the distribution of fermions (for instance in metals and neutron stars), in chemistry (in auto-catalytic reactions) and in other fields from oncology, economics to linguistics. Essential is *cooperativity or collective effects*[2] in the process (here, population change) being described.

One can say that t and f are now being measured in their natural units since the final form of equation (7.3) is universal – it does not depend on the details of any system: We can get back from general solutions $z(u)$ to any real population f in real time t by:

$$f(t) = \frac{a}{a_1}z(at).$$

The solution $z(u)$ to the universal equation translates into *any* population's development in time by scaling, if it obeys equation (7.2).

---

[2] for instance in the right hand side of equation (7.2), f weeds are consuming nutrient, and the amount of nutrient available is itself reduced by a_1f.

2. Features of the "universal" equation.

a) What do the factors $\dfrac{a}{a_1}$ and a do to axes of the graph of $z(u)$ when trans-
forming to $f(t)$?

b) Show that the population no longer increases without limit as time increases. What are the stable, long-term values of the plant population (in
terms of z and in terms of f)?

c) At what z is the maximum value of dz/du attained, and what is that maximal slope?

d) Evaluate the second derivative d^2z/du^2 and give its value at the z considered above. What is this point called?

e) From what you have assembled thus far, sketch the function $z(u)$.

3. Solving the logistic equation.

a) Solve the first order differential equation (7.3) for $z(u)$ by separation and
integration (you will need partial fractions). Use as a starting point $z = 1/2$
and a corresponding value, u_0, of u. [This is called an initial, or a boundary
condition. You need 1 for each order of the differential equation.] You may
find it convenient just to set $u_0 = 0$.
Rearrange any solution you get so that it looks like $z(u)$, rather than $u(z)$,
if that is what you ended up with. This function is called the Fermi–Dirac
function in quantum physics.

b) Confirm your solution conforms to the qualitative limits and special values you deduced above in Ex. 7.5.2.

c) The solution should display a "sigmoid" shape – confirm this from your
solution and your knowledge of the hyperbolic function that $u(z)$ can be
expressed as. This shape is the signature of collective effects. It is also seen
in biology, for instance in the uptake of oxygen by haemoglobin and essen-
tially is what allows oxygen transport in animals.

d) What about $u < 0$? Is your solution valid? Think also about the symmetry
$z \rightarrow 1 - z$.

Population regulated by a *predator*.

The population of weed above was limited by a finite source of nutrient in the pond of fixed size, rather than by predation. The so-called "predator-prey" equations govern the populations of two species where one depends on the other (for food). When prey is plentiful, predators will be better fed and their population will grow, with a lag behind that of the prey population. When predators have become plentiful, the prey will decline in numbers and, following that the predator population will decline. In turn the prey population will recover and the process continues cyclically.

Consider now the water weed predated upon by the grass carp (the white amur – *Ctenopharyngodon idella*). Now the weed population $f(t)$ is governed by:

$$\frac{df}{dt} = af - a_1 cf. \tag{7.4}$$

where the reduction in population term, $a_1 cf$, now says that f is reduced at a rate depending on f itself and on the number $c(t)$ of carp that are feeding on the weeds. Ignore limits on the weed population due to over-consumption of nutrient.

4. The carp themselves reproduce at a rate proportional to their own number. The proportionality is set by a constant b and by the weed population f; more weed supports more carp! Carp die at a rate $b_1 c$. Derive a differential equation for c analogous to equation (7.4) for the weed population f.

5. The predator-prey equations.
   Divide equations (7.4) and the carp equation derived above by a and absorb a into the d/dt as $(1/a)d/dt \to d/d(at) = d/du$, that is $u = at$ is a reduced (dimensionless) time as before.

   a) Show that this pair of equations become:

   $$\frac{dz}{du} = z - zy \tag{7.5}$$

   $$\frac{dy}{du} = zy - \beta y, \tag{7.6}$$

   where z and y are scaled versions of the weed and carp populations f and c respectively.
   Give an expression for β, the only parameter remaining after scaling. Give the scalings that take f, c to z, y.
   These are the Lotka–Volterra equations.

b) Confirm the equations give the qualitative limits and special values analogous to those you deduced above in Ex. 7.5.2.

c) Find the pairs of values of (z_0, y_0) that, if one could achieve them, would
not change further with time.

d) What happens to z if predators y are eliminated? What happens to y if
prey z runs out, and what then is the characteristic time for change in y?

The equations cannot be solved exactly for $z(t)$ and $y(t)$, though they can
be plotted easily. As anticipated, the predator population tracks that of prey.
We can, though, solve the equations for small deviations of (z, y) from (z_0, y_0),
and we can solve for $y(z)$:

6. Dynamics of predator-prey close to equilibrium.
Since we are looking for small excursions of the populations, write $z = \beta + p$
and $y = 1 + q$ where p and q are small. The product pq is therefore even
smaller and we neglect it.

a) Putting these forms for z and y into equations (7.5) and (7.6), give expressions for dp/du and dq/du.

b) Differentiate your result for dp/du and use your expression for dq/du to
get an equation for p only.

c) Thus find expressions for the excursions $p(t)$ and $q(t)$ from equilibrium.

d) Show how the amplitude P of the $p(t)$ variation is related to the amplitude Q for the variation of $q(t)$.

e) Sketch the variation of the fluctuations of the two populations away from
β and 1 in time, and say *how* one follows the other.[3]

f) Dividing $\frac{dy}{du}$ by $\frac{dz}{du}$, separating and integrating, show that y depends on z
via the relation of the form:

$$ye^{-y} = Cf(z, \beta), \tag{7.7}$$

where f is a function of z and β only, and C is a constant depending on the
initial choices z_i and y_i of z and y. You may have to manipulate logs etc. in
your answer to get it into the form given here. Give the form of f and of C.

---

[3] Python resources for this book, including for population dynamics, can be found at
https://isaacmaths.org/book/python_exercises

7.6   Parametric curves, circular coordinates, and vector calculus

Many curves or trajectories are specified by a parameter, for instance t, which governs position as $(x(t), y(t))$. The parameter might be time, arc length along the curve, or simply angle. The curve is a vector position $r(t) = (x(t), y(t))$ that advances in space as t increases – here it is 2-D, but it could be in 3-D: $r(t) = (x(t), y(t), z(t))$ or in higher dimensional, abstract spaces. Additional vectors associated with curves are for instance their tangents. The change of vectors, in time or space, that is their calculus, governs all mechanics in more than 1-D and most of fields (electromagnetism, gravitation etc.).

Circular motion

Changing direction at constant speed means accelerating perpendicular to the trajectory, towards the centre of local curvature. Steady motion in a circle, with centripetal acceleration to the centre of $a = v^2 / r$, is a familiar example. Such an acceleration requires a force and, when this is gravitational we have a celestial orbit.

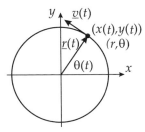

Figure 7.7. Cartesian and circular coordinates (x, y) and (r, θ) (radius r and angle θ) for position. Position, r, and velocity, v, vary with time, t, for a particle executing circular motion.

1. Draw your own version of figure 7.7. Take $\theta(t) = \omega t$ or $d\theta / dt = \omega$ for *uniform* circular motion (ω is the angular frequency and t is time). One can say that the position $(x(t), y(t))$ is given parametrically in terms of t.

   a) Write $r = (x(t), y(t))$ in terms of t, ω and of r (the magnitude of r).

b) Now expressing $x(t)$ and $y(t)$ as functions $g(t)$ and $h(t)$, find the differential equation obeyed by g and (separately) by h.

c) Confirm that the magnitude $|r|$ of the position vector is a constant in time, as expected for motion in a circle.

d) Find the velocity vector $v = \mathrm{d}r/\mathrm{d}t$ by differentiating the components of the time-varying position vector $r(t)$. This is the tangent vector to the curve $r(t)$.

e) What is the magnitude of v, that is $v = |v|$?

f) Give an expression for the vector dot product $v \cdot r$ (see vectors concept sheets isaacphysics.org/concepts/cm_vectors2)

g) Differentiate the components of v with respect to time to get the acceleration vector, a. Prove explicitly it is in the direction of $-r$ and give the constant of proportionality to get an expression for a in terms of ω and r.

2. Gravitational orbits

Circular motion involves acceleration towards the "centre". Consider a large body of mass M to provide this centripetal force (via gravity) for a small mass m that is in circular motion about it.

a) Write down Newton's gravitational (vector) force f of M on m where M is at the origin and m is at r. Remember the direction (sign) of f.

b) Derive the *vector* equation of motion, using Newton's Second Law, for m orbiting about the much larger M. Use the form of the acceleration that involves ω.

c) Derive Kepler's law that connects ω and r for the orbit.

d) Connect the period and radius of the orbit.

Tori

A torus can be thought of as two circles compounded with each other at right angles; see Fig. 7.8(a). They can be expressed parametrically, as above, but two parameters are needed – an angle θ for a circle in the x–y plane of radius R, and an angle α for a circle of radius a in the vertical plane and centred on the former circle. The circle of radius a is swept around as θ advances; see Fig. 7.8(b) for how the coordinates of a vector position $r(\theta, \alpha)$ on the surface of the torus are defined.

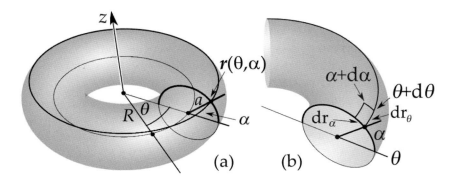

Figure 7.8. (a) A torus of interior radius R and of sectional radius a (here $a = 0.45R$). The coordinates θ and α define points on the torus. Circles in the vertical plane have, at fixed θ, the angle α varying in the interval $(0, 2\pi)$. At fixed α, varying $\theta \in (0, 2\pi)$ gives contours around the torus – one such is shown for $\alpha = \pi/4$. (b) Increments $d\theta$ and $d\alpha$ in angle cause increments dr in position in the surface. Differential vectors are denoted by dr_θ, along the θ circle for $\alpha = \pi/4$ fixed, and by dr_α, at constant θ.

3. Describing a torus

a) Give, in the form $(x(\theta), y(\theta), z(\theta))$, the parametric form of the circle of radius R, in the $z = 0$ plane, that forms the centre of the torus (see Fig. 7.8(a)).

b) The parametric equation of a point on the surface of a torus is of the form $r = (x(\theta, \alpha), y(\theta, \alpha), z(\theta, \alpha)) = (f(\alpha) \cos \theta, f(\alpha) \sin \theta, g(\alpha))$. By considering the displacement vector from the central circle to the surface at a point defined by α, give the functions $f(\alpha)$ and $g(\alpha)$. [They will clearly depend on R and a.]

Tangents and normals to surfaces.
Differential elements (vectors) dr_θ and dr_α of length in the surface are generated by incremental increases in θ and in α:

$$dr_\theta = \frac{\partial r}{\partial \theta} d\theta \quad \text{and} \quad dr_\alpha = \frac{\partial r}{\partial \alpha} d\alpha$$

where $\partial/\partial\theta$ is a "partial derivative": differentiate with respect to θ, in effect $d/d\theta$, while keeping other variables fixed, here α. See see Fig. 7.8(b) for the parallelogram formed by these two vectors advancing in the θ and α directions. These two vectors in fact define the tangents to the surface, pointing in the θ and α directions. These vectors trace out the surface as θ and α vary, and completely define it.

The vector product of these two vectors gives an area dA associated with advancing to $\theta + d\theta$, $\alpha + d\alpha$, that is $dA = dr_\theta \wedge dr_\alpha$, which is directed along the surface normal at this point. See section 7.2 for vector area and 6.1 for vector products. Concept sheets are available on-line on Isaac. Check that the vector dA points outward from the torus surface, the way it has been defined here.

4. Area and volume of a torus

a) Give, in the form $(x(\theta, \alpha), y(\theta, \alpha), z(\theta, \alpha))$, the parametric form of the differential tangent vectors (i) dr_θ and (ii) dr_α. [Take out factors of $d\theta$ and $d\alpha$ respectively.]

b) What is the scalar product $dr_\theta \cdot dr_\alpha$?

c) Give an expression for dA, the magnitude of dA.

d) Calculate the total surface area of this torus.

e) Calculate the volume enclosed by the surface of this torus. [Hint: Find the differential element of volume associated with incrementing separately by $d\theta$, by $d\alpha$ and by da' (along a radius of a sectional circle. Integrate $\int_0^a da'$ over the radius of a section.]

A marvellous spiral
The "spira mirabilis" of Jacob Bernoulli has remarkable properties:

1. As it spirals out from the origin, the log spiral makes a constant angle, ϕ, with radius vectors. See figure 7.9(a).

2. Subsequent crossings of a given radius have their distances from the origin in constant ratio to each other. This means the distance of the curve from the origin grows geometrically, that is exponentially. These spirals are thus self-similar curves.

3. These spirals can be given parametrically in circular polar coordinates, using angle θ or arc length along the spiral, s, as parameters.

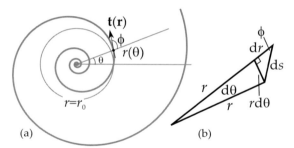

Figure 7.9. (a) A logarithmic spiral $r(\theta)$ parameterised by the angular position θ: The spiral (tangent vector $t(\theta)$) makes a constant angle ϕ with radii. Here, $\phi = 1.4\,\text{rad} \equiv 80.2°$. (b) Differential triangle connecting increments in r, θ and spiral arc length, s.

We take at $\theta = 0$, the radius to be $r = r_0$.

5. The logarithmic spiral

a) Considering the differential triangle of Fig. 7.9(b), find the relationship between $(ds)^2$ and the other infinitesimals $(dr, d\theta)$.

b) Show that $dr/d\theta = br$ and give an expression for the constant b.

c) Give the equation of the spiral in the form $r(\theta)$, recalling any boundary conditions given.

d) For this spiral, what is the ratio of radial distances out to successive crossings a given radial line, that is, at a given θ (or $\theta + 2\pi n$ for later crossings of the same radius)?

e) Considering $ds/d\theta$, find an expression for $s(\theta)$.

f) The spiral has an infinite number of turns to get to the origin $r = 0$. What is the arc length L along the spiral from $r = 0$ to r_0?

g) Give the equation of the spiral in its parametric form $(r(s), \theta(s))$. [This form is known as a "unit speed parametrisation" in differential geometry since the advancement of s corresponds to geometric distance s along the curve.]

h) A "golden spiral" is a logarithmic spiral where the coordinate r increases by a factor of the golden ratio, $(1 + \sqrt{5})/2$, every quarter turn (θ advancing by $\pi/2$). It is the limit of the spiral created by quarter arcs of circles in squares increasing in size as the Fibonacci sequence. Calculate the golden angle ϕ_g for such a spiral.

Logarithmic horns

Logarithmic spirals occur in nature in the growth and form of organisms, for instance the spirals of plant cones can sometimes relate to the Fibonacci example above. In 3-D, the shell of the nautilus is an example; see Fig. 7.10(a) and (b). A horn that spirals and opens out in a logarithmic way can be constructed from log spirals and circles:

Consider the radial vector $(r(\theta)\cos\theta, r(\theta)\sin\theta, 0)$ where the last entry 0 says there is no z component to this vector, and $r(\theta)$ was derived above for log spirals (with a given b and r_0). Check that the vector's length is simply $r(\theta)$. This is the position of a point, defined by θ, on the central spiral of the horn, Fig. 7.10(c). We now want to associate a sectional circle with this point, as we did with the torus. It is in the plane $\theta = $ constant and r and z varying (the sectional cut in Fig. 7.10(c); note that the circle is not perpendicular to the spiral at its centre). Let the radius w of this circle increase as we increase θ so it forms the surface of a horn.

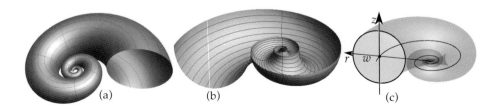

(a)   (b)   (c)

Figure 7.10. (a) A logarithmic horn with a gap between the surfaces of successive turns. See the book cover for a colour version. It has $b = 0.2$ and $w = 0.5$ (see text). (b) A logarithmic horn where successive turns touch. It has been cut in its central plane to reveal the interior. (c) A circular section, of radius w, of a horn that has been cut by the r–z vertical plane of $\theta = $ constant. The circle is centred at $r(\theta)$, that is, on the central, log spiral (black line) that intersects the plane at the dot.

a) Give the parametric form of a circle of radius w centred on the point at $(r(\theta)\cos\theta, r(\theta)\sin\theta, 0)$, and in the vertical plane. That is, give the parametric form of vectors from the origin in Fig. 7.10(c) to points on the circle. As in question 7.6.1, you will need an angle, α say, that parametrically describes points on the sectional circle in Fig. 7.10(c). [Take $((r(\theta)+w)\cos\theta, (r(\theta)+w)\sin\theta, 0)$ to correspond to the point on the circle with $\alpha = 0$.]

b) If $r(\theta) = r_0 f(\theta)$, where $f(\theta)$ has been determined above (the log spiral, question 7.6.5(c)), let the sectional circle centred on radial position $r(\theta)$ equally have its radius scaled by the same $f(\theta)$, that is $w(\theta) = w_0 f(\theta)$, so that it grows in proportion to the spiral. What condition is there on w_0 in terms of b so that the surface of the horn centred on $r(\theta)$ just touches that centred on $r(\theta - 2\pi)$, that is it touches on the part radially inside? The just-touching case is shown in Fig. 7.10(b), whereas that in (a) has $b = 0.2$ and $w_0 = 0.5$, with $r_0 = 1$. Check that this horn does indeed conform with your condition.

c) What is the overall parametric vector formula for logarithmic horns in terms of the variables θ and α, and the constants r_0, b and w_0? The answer is a Cartesian vector of the form $r = (x(\theta,\alpha), y(\theta,\alpha), z(\theta,\alpha))$.

7. Some differential geometry of logarithmic horns

a) Show that

$$d\mathbf{r}_\theta = [br + f(\theta)((-(r_0 + w_0\cos\alpha)\sin\theta, (r_0 + w_0\cos\alpha)\cos\theta, 0))]\,d\theta$$

Give, also in the parametric form $(x(\theta,\alpha), y(\theta,\alpha), z(\theta,\alpha))$, the differential tangent vector $d\mathbf{r}_\alpha$, analogously to in question 7.6.4(a).

For ease of calculation, now consider horns with $w_0 = r_0$. Sectional circles grow exactly as the spiral radius does so that the inner edge of the horn is always at the origin; see Fig. 7.11.

b) Show that $d\mathbf{r}_\alpha$ is perpendicular to the second part of $d\mathbf{r}_\theta$. Evaluate the scalar product $d\mathbf{r}_\theta \cdot d\mathbf{r}_\alpha$. Discuss why the answer is non-zero. [Not limited to this special case.] What component of the variation of r with θ makes the differentials non-perpendicular?

c) Explain why $|dA|^2 = |d\mathbf{r}_\theta \wedge d\mathbf{r}_\alpha|^2 = |d\mathbf{r}_\theta|^2 |d\mathbf{r}_\alpha|^2 - (d\mathbf{r}_\theta \cdot d\mathbf{r}_\alpha)^2$.

d) Hence find dA, the magnitude of dA.

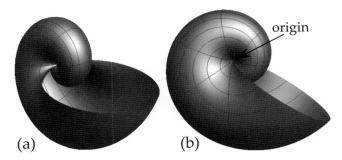

Figure 7.11. A horn where the radius of the circular section matches the spiral radius; $w_0 = r_0$. The spiral is (a) reentrant, and (b) also always touches the origin.

e) Calculate the total surface area of this horn for $\theta \in (-\infty, 0)$. Recall that at each section, the horn has an inner and outer surface, of equal magnitudes.

8. Miscellaneous geometry problems – on-line resources:

a) Calculate $d\mathbf{r}_\theta \wedge d\mathbf{r}_\alpha$ the direction of the vector area $d\mathbf{A}$. The unit vector from this is the surface normal at the point defined by θ, α.

b) *Calculate the volume enclosed by the surface of this horn for $\theta \in (-\infty, 0)$. The guidance given for tori for calculating their volume will be relevant. It is in error only at second order in $d\theta$ and is hence accurate in the limit.

c) Using Python, or another language, draw spirals and horns.
[Python scripts and guidance to get started, including for spirals and shells, are available on Isaac on-line.[4]]

d) Contrast with the totally different Archimedean Spiral.
[The Python exercises are helpful with these spirals too.]

---

[4]See https://isaacmaths.org/book/python_exercises

7.7   Wind–driven yachts, sand yachts and ice boats

We explore sailing across the wind where sails act as aerofoils, very much as aircraft wings give lift. How do wind–driven craft travel faster than the wind? We break the question into sub-questions:
How is the propulsion force generated from the wind?
What is the maximum speed that a wind–driven vehicle can attain? At what angle must you sail to the wind to attain this maximum speed?
What limits the actual speed of a wind–driven craft?

Fluxes and Forces

The aerofoil offers lift (and suffers some drag) when it harnesses the momentum flux of the wind flowing around it; see Figure 7.12(a).

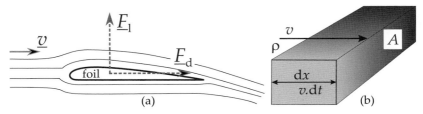

Figure 7.12. (a) An aerofoil section with the lift and drag forces, $\underline{F}_l$ and $\underline{F}_d$. Flow is deflected downwards. The overall momentum change of the flow and forces on the wing add up. (b) Fluid of density ρ moves with speed v. The cuboid of fluid of end area A and thickness $dx = vdt$ passes through A in time dt.

Momentum *flux density* is the rate of flow of momentum intercepted per unit area, see section 7.2, and is a force per unit area; see Figure 7.12(b). The cuboid of fluid passing through the area A in time dt has volume $dV = Avdt$ and therefore mass $dm = \rho Avdt$, where v is the wind speed, and ρ is the fluid density. Its momentum is:

$$dp = vdm = A\rho v^2 dt \quad \rightarrow \quad \frac{dp}{dt} = A\rho v^2. \tag{7.8}$$

Recall Newton's second law: force is the rate of change of momentum. The momentum *flux density* (that is, momentum per unit time per unit area), on dividing momentum by dt and A, is then ρv^2. Forces from such flows, from dp/dt, are of the form $A\rho v^2$. The drag and lift forces[5] on an *aerofoil* depend on the momentum flux, with directions as in Figure 7.12(a):

$$F_l = \tfrac{1}{2}c_l A_f \rho_a v^2 \qquad F_d = \tfrac{1}{2}c_d A_f \rho_a v^2. \qquad (7.9)$$

A_f is the frontal projected area of the foil, c_l and c_d are its lift and drag coefficients, the $\tfrac{1}{2}$ is conventional, v is the wind speed *seen by the foil* — vital in what follows — and ρ_a is the density of air.

1. Add an arrow for the resultant force, $\underline{F}$, in Figure 7.12(a).

    a) Give an expression for γ, the angle $\underline{F}$ makes with $\underline{F}_l$, in terms of F_l and F_d.

    b) For a yacht with values[6] of $c_l = 2.5$ and $c_d = 1.0$, determine a value for the angle γ. Clearly the smaller the drag force is compared with the lift force, the smaller γ is (a desirable characteristic).

The "apparent wind" direction on a moving vehicle.

 Consider wind with velocity $\underline{w}$ and a wind–driven craft, such as a yacht, proceeding with velocity $\underline{u}$ (Figure 7.13(a)). As you sail, the "apparent wind" velocity $\underline{v}$ gets closer to the line of your direction of motion, that is $\beta < \alpha$. The aerofoil is aligned around the direction $\underline{v}$ to generate the forces discussed above. Masts often have a pennant at their tip; it points in the direction of $\underline{v}$. For a given α, one can then align[7] the sail close to $\underline{v}$. This apparent velocity changes direction due to changes in speed u or direction α.

2. In terms of u, w and α, find expressions for:

    a) $v \cos \beta$

    b) $v \sin \beta$

---

[5]Can you see why these two forces are sufficient to describe the aerodynamic force on the wing?

[6]The lift coefficient determines the wing's efficiency: 2 to 2.5 for a yacht aerofoil, 1.5 to 2 for a traditional-style sail.

[7]Choosing the precise alignment, and the shape of the sail, is the skill of the sailor, affecting c_l, c_d and hence γ, and differing according to α – something ignored here.

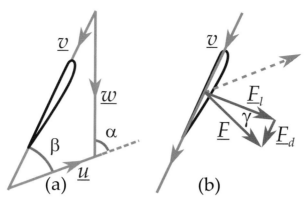

Figure 7.13. (a) Motion $\underline{u}$ in a wind $\underline{w}$ gives an apparent incident wind velocity of $\underline{v} = \underline{w} - \underline{u}$ at an angle β to $\underline{u}$. We simplify by aligning the aerofoil in the direction of $\underline{v}$. (b) Wind $\underline{v}$ and forces $\underline{F_l}$ and $\underline{F_d}$, perpendicular and parallel to $\underline{v}$ respectively, drawn in the frame of the boat. We assume that motion perpendicular to the direction being steered (dotted) is suppressed - e.g. by a keel or centreboard.

c) Find F_u, the aerofoil's force on the yacht along the direction of its motion, in terms of F, β and γ, where $F^2 = F_d^2 + F_l^2$.
[Hint: See Figure 7.13(b); the force F includes elements of lift and drag.]

d) Figure 7.13(b) is in the frame of the craft (the frame of reference in which the craft is at rest). "Stall" is when the force, in the direction of $\underline{u}$, drops to zero when at zero speed. Give an expression for α_{min}, the closest one can sail to the wind before approaching a stall.

The maximum speed of a craft.
A craft will reach its maximum velocity when the *net* force on it along $\underline{u}$ is zero: $F_u = F_f$ where F_f is the resistive force, other than from the sail, impeding the motion.

3. a) Show that the force along the direction of travel is of the form:

$$F_u = \left(\tfrac{1}{2}\rho_a w^2 Ac\right) \frac{v}{w} f\left(\frac{u}{w}, \alpha, \gamma\right) \qquad (7.10)$$

where $c = \sqrt{c_l^2 + c_d^2}$, and give an expression for the function f. [Hint: use an expression for F_u from above.]

The pre-factor is the scale of force from simply being in the wind, $F_w = \frac{1}{2}\rho_a w^2 A c$. The ratio F_u / F_w measures how effective the forward propulsion force F_u actually is – a figure of merit. In physics such a ratio is known as a reduced quantity. Another such is $\frac{u}{w}$, the speed of the craft divided by the wind speed. Reduced quantities are dimensionless, i.e. without unit.

b) The reduced apparent wind speed is $\frac{v}{w}$. Give an expression for it in terms of $\frac{u}{w}$ and the angle α of motion with respect to the wind.

c) Consider the case where there is no resistance to motion, other than F_d from the sail itself. Show that the maximum reduced speed is then of the form $\frac{u}{w}|_{max} = g(\alpha, \gamma)$, and give $g(\alpha, \gamma)$.

d) Find the angle of motion, α_{max}, at which $\frac{u}{w}|_{max}$ is itself maximised.

Determining the actual speed, u, of Sand Yachts and Ice Boats.
Speed is set by force balance, that is, when propulsion and resistive forces are equal: $F_u = F_f$. An ice boat and a sand yacht are both wind-driven craft; the ice boat sits on runners and the sand yacht on wheels. Both crafts experience resistive forces caused by friction given by $F_f = \mu_k mg$ where μ_k is the coefficient of kinetic friction and m their mass; see Figure 7.14(a), and also Figure 7.14(b) for plots of F_f / F_w with various μ_k. This type of resistance is independent of u.

4. a) Figure 7.14 shows F_u / F_w and F_f / F_w against $\frac{u}{w}$ for various degrees of friction. For $F_f / F_w = 0.9$, determine whether point 1 or point 2 of $F_u = F_f$ is stable if u/w were to deviate slightly, either way, away from the intersection value. Such fluctuations must happen all the time while in motion. Therefore these considerations determine at what speed the craft stably travels. What about the character of intersection 3 when friction is such that $F_f / F_w = 0.6$?

b) What is F_u / F_w at $u = 0$ for the values $\gamma = 10°$ and $\alpha = 60°$ of Fig. 7.14(b)?

c) What is the maximum possible reduced speed, $\frac{u}{w}|_{max}$, of this ice boat? Ignore frictional forces and use values of $\gamma = 10°$ and $\alpha = 60°$, as in Figure 7.14. You have thus located point 5 on this graph. This maximal u/w value is not much greater than the real case, point 4.

d) For each friction line, *estimate* the reduced speed that needs to be reached before the craft moves under the force of the wind only. [Hint: See Figure 7.14(b). For the case $F_f / F_w = 0.9$ you will need to establish the u/w

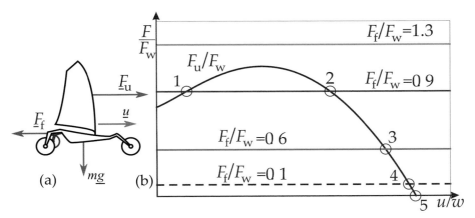

Figure 7.14. (a) The weight mg, the resistive force $F_f = \mu_k mg$ and the driving force F_u acting on a sand yacht. (b) The aerofoil force along direction of travel, F_u/F_w, and frictional force, F_f/F_w, (both divided by the force scale F_w of the wind), plotted against speed of the craft u/w (divided by the wind speed). The angle $\gamma = 10°$ is that of a typical sand yacht, and the angle of travel to the wind is $\alpha = 60°$. The horizontal lines of F_f/F_w are typical for a sand yacht. The dotted line is a typical value of F_f/F_w for an ice boat.

axis scale from the $F_u/F_w = 0$ intercept, point 5, and use measurement along that axis.]

e) What is the smallest value of the reduced frictional force F_f/F_w for which no motion is possible? [Hint: use your answer in (b) which sets the vertical scale of the diagram, for the γ and α values of Figure 7.14(b), and measure with a ruler along the F/F_w axis.]

f) *Estimate* the maximum reduced speed of the ice boat, without ignoring frictional forces, for $F_f/F_w = 0.1$. [Hint: your answer in part (d) sets the diagram's horizontal scale.]

g) Estimate the reduced speed of the sand yacht at point 3.

Determining the actual speed, u, of water yachts.
For a normal, water yacht, the resistive forces $F_f = \frac{1}{2}\rho_w u^2 A_b c_f$ are caused rather by the inertial drag from moving through the water at speed u with respect to the water, A_b being the sectional frontal area the boat presents to the water, c_f the drag coefficient, and ρ_w the density of water; see Fig-

ure 7.15(a). The wind force scale $F_w = \frac{1}{2}\rho_a w^2 A c$ is still the measure of force, see Figure 7.15(b). Note that now the angle γ is large enough to make F_u a simple decreasing function of u/w.

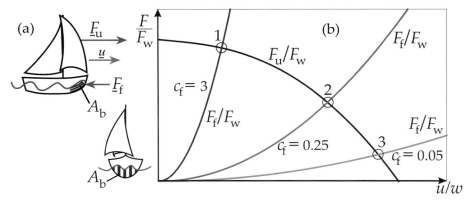

Figure 7.15. (a) Inertial drag F_f from the water, and the driving force F_u for a water yacht. (b) Propulsion force divided by the scale of wind force, F_u/F_w, for γ based on $c_l = 2.5$ and $c_d = 1.0$, and for $\alpha = 60°$, and inertial drag F_f/F_w likewise reduced and shown for marked c_f, against u/w.

5. a) Find the slope of F_u/F_w at $u/w = 0$.

b) In Figure 7.15(b), determine whether points 1, 2 and 3 are stable points of the motion, as in question 4.a.

c) What is the maximum possible reduced speed[8], u/w, of this yacht? Ignore inertial drag from the water and use values of γ calculated from c_l and c_d and $\alpha = 60°$ in Figure 7.15(b).

d) Assuming the lowest given inertial drag F_f/F_w, *estimate* the maximum realistic reduced speed of the yacht. [This very low c_f is probably for a hydrofoil yacht.]

---

[8]In late 2012 the Vestas Sailrocket 2 skippered by Paul Larsen achieved a new outright world speed record of around 2.5 times the speed of the wind. Also in 2009, the world land speed record for a wind-powered vehicle was set by the sand yacht Greenbird sailing at about three times the speed of the wind. On ice it is accepted that craft often travel at as much as five times the speed of the wind. Compare these values with your calculations.

7.8   Rays, rainbows, and caustics

A rainbow is one of the most beautiful of all physical phenomena; figure 7.16(a). They arise when, with the Sun behind you, light rays are
deflected by rain drops to your eyes; figure 7.16(b).

Why do rays arrive at your eye in arcs with such sharply defined angle,
and what is that angle? Why do rays of different colour subtend slightly
different angles to your eye? Why do doubles occur?

Rays and geometry

Consider a spherical droplet (refractive index n) with parallel rays from
the Sun incident on it. Depending on where the beam strikes the sphere,
the angle of incidence α varies. A beam refracts into the drop, internally
reflects and then refracts out; see figure 7.16(b) on which some radii have
been drawn. Draw a diagram showing the direction of the Sun, droplets in a

Figure 7.16. (a) Mark Seton: "Double Rainbow over Doctors Pond".
(b) A ray passing through a droplet. All drops in an arc of angle χ to
rays from the Sun have the same path of light rays through them.

range of positions, your eye, and the subtended angle χ of the arc of droplets
which share this incidence angle, α.

1. a) Find the angle χ the exit beam makes with the direction of the incident
   beam in terms of angles α and β. [Hint: find angles in the drop. Add *deflec-
   tions* of the beam at A, B & C.]

In fact χ is purely as a function of α, since Snell's law, $\sin(\beta) = \sin(\alpha)/n$, gives $\beta(\alpha)$ in χ. Figure 7.17(a) plots $\chi(\alpha)$ against α showing, for suitable refractive index, there can be a maximal deflection χ_m at an optimal angle of incidence, α_m:

b) At the maximal $\chi(\alpha, \beta)$ find the rate of change of $\beta(\alpha)$ with α.

c) Given β and α are related by Snell's law, and given the above connection between the optimal β and α, find an expression for $\sin \alpha_m$ in terms of n.

d) Find the corresponding expression for $\sin \beta_m$.

e) Find χ_m purely in terms of n.

f) What is χ_m for water droplets ($n = 4/3$)?

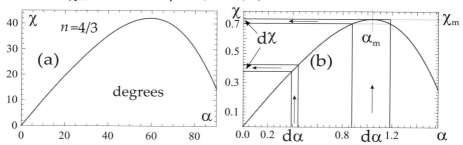

Figure 7.17. (a) Deflection $\chi(\alpha)$ against incidence α (degrees) for water. (b) Deflection $\chi(\alpha)$ against incidence α (radians) for water. Angular intervals of α are mapped into those of χ. The ratio $d\alpha/d\chi$ becomes critical at maximal deflection.

Why are rainbows bright?

Refraction at A and C is accompanied by some external/internal reflection. Internal reflection at B has associated refraction of light out of the drop. The weight of reflection versus refraction depends on the "Fresnel Coefficients". They are functions of α at A and of β (and hence, through Snell's law, actually of α) at B and C. The emergent light intensity $H(\chi)$ times the angular interval $d\chi$ at C, that is $H(\chi)d\chi$, depends on the intensity initially striking the drop at A, in the interval $d\alpha$, multiplied successively by the three relevant Fresnel Coefficients, compounded together as $I(\alpha)d\alpha$ say. These expressions are equal:

$$H(\chi)d\chi = I(\alpha)d\alpha \quad \text{or} \quad \frac{H(\chi)}{I(\alpha)} = \frac{d\alpha}{d\chi} \tag{7.11}$$

The angular distribution of intensity $H(\chi)$ that we see is very different from the input $I(\alpha)$. The derivative $d\alpha/d\chi$ diverges at (χ_m, α_m): The intensity $I(\alpha)$ in an interval $(\alpha, \alpha + d\alpha)$ is mapped into a much smaller interval $(\chi, \chi + d\chi)$, that is, the light is intensified, $H(\chi)/I(\alpha) \to \infty$, and this arc of intensification is the rainbow – see Figure 7.17(b).

2. a) Why is the sky under a rainbow brighter than that above?

b) What happens to the rainbow in the limit the refractive index of a liquid vanishes, $n \to 1$?

c) What happens when n becomes very high (these are toxic, arsenic-based liquids — do not try at home!)? Give the limiting refractive index for rainbows.

Colours, shape, other liquids, orders

3. a) A fit for how the refractive index of water changes with wavelength λ of light in the visible region is:

$$n(\lambda) = 1.31848 + 6.662/(\lambda[\text{nm}] - 129.2) \qquad (7.12)$$

where λ is given in nanometers (blue $\approx$ 450, green $\approx$ 530, red $\approx$ 630). The variation $n(\lambda)$ is called "dispersion". Thus different wavelengths of light have different angles χ_m where they are brightest, and rainbows get their colour. Calculate the angular width of a rainbow between blue and red.

b) Why is a rainbow darkest just outside the blue band?

c) On the moon Titan of Jupiter it rains liquid methane ($n = 1.286$). What is the rainbow angle, χ_m, there?

d) Calculate the rainbow angle if it were to rain liquid hydrogen (with a very low refractive index $n = 1.0974$). What qualitative features would the rainbow angle have? How would you observe it?

e) We have considered a drop in the sky at an angle χ_m to the direction of the Sun's rays. Drops anywhere on a circle subtending χ_m to the eye with the Sun shining perpendicular to the circle will contribute equally (hence the "bow"). But some drops might be below the horizon! What ways are there of seeing complete rainbows?

f) How does the higher order bow arise? [Hint: you have to re-do the consid-erations of figure 7.16(b) from scratch; there are more internal reflections.] Calculate its subtended angle for a given n. Discuss why it is not as bright as the first order.

g) What was the elevation of the Sun in the sky when Mark Seton took his photo (see Figure 7.16(a))? [Hint: the secondary rainbow helps.]

7.9   Dr. Conduit's 101 Integrals

These integrals range in difficulty from A-level to difficult, but all rely on school methods. If you do call upon help, be sure to execute the integral yourself too. Familiarity with the method will allow you to adapt it to other integrals. Seeing such a range of integrals will give experience in picking the method needed, and then it is a matter of application of the tools you have.

Hex. ↓	(a)	(b)	(c)
7.9.1	$\int x \sec^2 x \, dx$	$\int \ln \sqrt{1 + x^2} \, dx$	$\int \sin(\ln(x)) \, dx$
7.9.2	$\int \dfrac{x^3}{\sqrt{16 - x^2}} \, dx$	$\int \dfrac{8}{x^4 + 2x^3} \, dx$	$\int \sin x \cos 2x \, dx$
7.9.3	$\int \dfrac{1}{(25 + 9x^2)^{5/2}} \, dx$	$\int e^{\cos x} \sin x \, dx$	$\int \dfrac{\cos x}{(5 + \sin x)^2} \, dx$
7.9.4	$\int \dfrac{\cos x}{\sqrt{1 + \sin x}} \, dx$	$\int x \sqrt{1 + x} \, dx$	$\int (\ln x)^2 \, dx$
7.9.5	$\int \dfrac{1}{x^4 + 4x^2 + 3} \, dx$	$\int \dfrac{1}{x(2 + \ln x)} \, dx$	$\int x^3 \sin x^2 \, dx$
7.9.6	$\int \sin^4 x \cos^3 x \, dx$	$\int \tan^2 x \sec^2 x \, dx$	$\int x \sec^{-1} x \, dx$
7.9.7	$\int \dfrac{\sqrt{1 - x^2}}{x} \, dx$	$\int \sin^4 x \, dx$	$\int \sqrt{9 - 25x^2} \, dx$
7.9.8	$\int \sec^3 x \, dx$	$\int x \sin x \, dx$	$\int \sin x \ln|\sin x| \, dx$
7.9.9	$\int \dfrac{1}{x^2 - 5x + 6} \, dx$	$\int \dfrac{1}{x^4 - 16} \, dx$	$\int \sin^5 x \cos^4 x \, dx$
7.9.10	$\int e^x \cos 2x \, dx$	$\int \dfrac{\tan x}{\sqrt{1 + \cos 2x}} \, dx$	$\int x \ln \sqrt{x + 2} \, dx$

Hex. ↓	(a)	(b)	(c)
7.9.11	$\int \dfrac{\cos \sqrt{x}}{\sqrt{x}}\,dx$	$\int \tan^{-1}(7x)\,dx$	$\int \sin 2x \cos 5x\,dx$
7.9.12	$\int \dfrac{1}{\sqrt{6x - x^2}}\,dx$	$\int \sqrt{x^2 - 9}\,dx$	$\int \dfrac{\sin x}{2 + \cos x}\,dx$
7.9.13	$\int \dfrac{1}{\sqrt{36 + x^2}}\,dx$	$\int \dfrac{\tan x}{\cos^2 x}\,dx$	$\int \dfrac{x + 1}{x^2(x - 1)}\,dx$
7.9.14	$\int \dfrac{1}{x^2 + 4x + 8}\,dx$	$\int \dfrac{x^5}{\sqrt{1 + x^2}}\,dx$	$\int \dfrac{\sqrt{x^2 - 5}}{x}\,dx$
7.9.15	$\int \dfrac{\sqrt{1 - x^2}}{x^2}\,dx$	$\int \dfrac{1}{(x^2 + 9)^2}\,dx$	$\int \dfrac{1}{(x^2 - 1)^{3/2}}\,dx$
7.9.16	$\int \dfrac{\sin^3 x}{\cos^2 x + 1}\,dx$	$\int (e^x + x^2)^2\,dx$	$\int \dfrac{\tan x}{\sec x - 1}\,dx$
7.9.17	$\int \dfrac{4x^5 - 1}{(x^5 + x + 1)^2}\,dx$	$\int \dfrac{x}{x^4 - 16}\,dx$	$\int \dfrac{\cot x}{\ln(e \sin x)}\,dx$
7.9.18	$\int x^4(1 + x^5)^3\,dx$	$\int \dfrac{1}{1 + \sin x}\,dx$	$\int \dfrac{\cos^4 x}{\sin x}\,dx$
7.9.19	$\int \tan^3 3x \sec 3x\,dx$	$\int \dfrac{1}{\sqrt{(3^2 + x^2)^3}}\,dx$	$\int \dfrac{\sin^{-1} x}{\sqrt{1 - x^2}}\,dx$
7.9.20	$\int \sin^{-1} x\,dx$	$\int \dfrac{1}{x\sqrt{x^2 - 1}}\,dx$	$\int \sec^6 x \tan^8 x\,dx$

Hex. ↓	(a)	(b)	(c)
7.9.21	$\int x^{1/4} \ln x\,dx$	$\int \dfrac{1}{\sqrt{x^2 - 2x}}\,dx$	$\int \dfrac{2x^3 + 3}{(x^2 + 1)^3}\,dx$
7.9.22	$\int \dfrac{1}{\sqrt{5x - 9}}\,dx$	$\int \dfrac{1}{\sqrt{e^{2x} - 1}}\,dx$	$\int \dfrac{x^9}{\sqrt{1 - x^5}}\,dx$
7.9.23	$\int (x^2 + 1) \ln x\,dx$	$\int \sin^3 x \cos^7 x\,dx$	$\int \dfrac{1}{\pi^2 + x^2}\,dx$
7.9.24	$\int \dfrac{5x - 1}{x^2 - x - 2}\,dx$	$\int \sec^5 x \tan^3 x\,dx$	$\int e^{2x} \cos 3x\,dx$
7.9.25	$\int x \tan^{-1} x\,dx$	$\int \dfrac{1}{\sqrt{4 - x^2}}\,dx$	$\int \sqrt{x^2 + 2x}\,dx$
7.9.26	$\int \dfrac{x^3 + x^2}{x^2 + x - 2}\,dx$	$\int e^x \sec^6(e^x)\,dx$	$\int \dfrac{1}{\sqrt{x}(1 + \sqrt{x})^2}\,dx$
7.9.27	$\int x^2 \cos 3x\,dx$	$\int \dfrac{1}{x\sqrt{9 + 4x^2}}\,dx$	$\int \dfrac{4x + 1}{x^3 + 4x}\,dx$
7.9.28	$\int \dfrac{\cos^7 x}{\sqrt{\sin x}}\,dx$	$\int \tan^2 x \sec x\,dx$	$\int x^3 \sqrt{1 - x^2}\,dx$
7.9.29	$\int \dfrac{\cos^2 x}{\sin x}\,dx$	$\int \dfrac{x + 1}{x^4 - x^3 - 20x^2}\,dx$	$\int \csc^8 x \cot^2 x\,dx$
7.9.30	$\int \dfrac{\sin^2(\ln x)}{x}\,dx$	$\int \dfrac{1}{\cos^2 x + 4\sin^2 x}\,dx$	

Hex. ↓	(a)	(b)
7.9.31	$\int x^{2/3}(x^{5/3}+1)^{2/3}dx$	$\int \dfrac{\csc^2 x}{(\csc x + \cot x)^{9/2}}dx$
7.9.32	$\int \dfrac{1}{x(\ln x - 1)\ln x}dx$	$\int \dfrac{5x^3 - 17x^2 + 19x - 13}{(x+1)(x-2)^3}dx$
7.9.33	$\int \dfrac{12x^4 + 190x^2 + 13x - 6}{(2x-1)(x^2+16)}dx$	$\int \dfrac{2x^2 - 3x + 2}{2x^3 + 11x^2 + 5x}dx$
7.9.34	$\int \dfrac{\sec^2 x}{\sec^2 x - 3\tan x - 1}dx$	$\int \dfrac{4x^3 - 7x^2 + 31x - 38}{x^4 + 13x^2 + 36}dx$
7.9.35	$\int 7x\sin^5(x^2)\cos(x^2)dx$	$\int x\sqrt{1-x^2}\sin^{-1}x\,dx$
7.9.36	$\int \dfrac{3x-7}{(x-1)(x-2)(x-3)}dx$	$\int \sec^7 2x\tan^5 2x\,dx$